Working
TECHNOLOGY

GERALD LLOYD
AND JOHN TRICKETT

CAMBRIDGE
UNIVERSITY PRESS

Published by the Press Syndicate of the University of Cambridge
The Pitt Building, Trumpington Street, Cambridge CB2 1RP
40 West 20th Street, New York, NY 10011–4211, USA
10 Stamford Road, Oakleigh, Victoria 3166, Australia

First published 1992

Printed in Great Britain by Ebenezer Baylis & Son Limited,
The Trinity Press, Worcester and London

A catalogue record for this book is available from the British Library

ISBN 0 521 42341 4 paperback

GE

Introduction

WORK-RELATED ASSIGNMENTS

'Education for economic and industrial understanding' is one of the cross-curricular themes identified in NCC Circular 6: *The National Curriculum and Whole Curriculum Planning* (November 1989). This text is an attempt to provide material that will help teachers of Design and Technology to cover this theme and the business and industry contexts within Technology in the National Curriculum. Included is a collection of ten pairs of 'work-related' Design and Technology assignments that have been prepared for Years 10 and 11 (i.e. 14 to 16-year-olds) in schools. Each is structured in such a way that it can provide an item of coursework assessment for the GCSE examination in schools and colleges.

Other collections of work-related assignments are available in the following curriculum areas: English, Mathematics, Science and Geography.

CONTENTS

NATIONAL CURRICULUM

The following table gives brief details of which units cover which Attainment Targets of the National Curriculum's Design and Technology scheme.

	Attainment Targets 1						2						3						4					
Employers	5	6	7	8	9	10	5	6	7	8	9	10	5	6	7	8	9	10	5	6	7	8	9	10
Blue Circle Cement	b		a				e	a					b d						b	a				
British Airways	b	c					c d							a			b		a	c				
British Rail		b c					d			b					c	a			c			a		
British Telecom			a	c						c	a			b		c			c				a	
Gibbs Mew Breweries		a	c					a	b				b	b					b	b				
Marks & Spencer			c	b			d e							b e						c d				
Royal Mail Parcelforce			d		b		a	b						e e						a	a			
University of Bath			c a				c		c					a b						a d				
Wellcome Foundation	b	c						b	b				a			c			b		a			
Wilton Royal		c				a		f	c					f	a						a	b		

MATERIALS

Each unit relates to the activities of a particular employer and contains a Teacher's Guide and Student Materials.

The Teacher's Guide outlines the students' assignments and provides more details of how the units relate to the National Curriculum.

The Student Materials contain background information about the company and its activities, and set two assignments based upon these activities.

TEACHING STYLE

During trials various teaching approaches were suggested. You will have to decide how prescriptive a style you wish to use.

Background information

In an attempt to give a real industrial/commercial context the description of each industrial activity is presented in the words of an employee of that company.

Trial teachers' suggestions for this information included:

(a) use it independently of assignments

(b) use it to gain background knowledge for a talk to give to students

(c) give it to students without the assignments with the requirement that they invent for themselves an assignment on which they would like to work

(d) give it to students with one or both assignments.

Assignments

There are two assignments in every unit, with each taking a different length of time for completion. The assignments can normally be studied independently.

The requirements for GCSE examination syllabuses vary considerably. Teachers, guided by their syllabuses, have to decide upon the most suitable assignment(s) for their students. Trial teachers' recommendations included the following:

(a) use them as a guide for producing your own assignments

(b) use them in modified form

(c) use them as they are, allowing students to work individually or in small groups.

Groupings

A recommendation of some trial teachers was that all students should work separately on the same assignment at the same time, while others recommended that students should all work separately at the same time on different assignments. All possible groupings were recommended, ranging from individuals and small groups to complete teaching groups. Some teachers found the role play occasionally required of them somewhat demanding.

You must decide upon the style of teaching that is most appropriate for your teaching group and the examination for which your students are to be entered, and – perhaps most importantly – one with which you will feel comfortable.

All Student Materials can be photocopied free of charge.

THE PROJECT

The production of these texts has represented a joint effort between industry, schools and local education authorities, and the University of Bath School of Education. The project has been funded by the Clerical Medical Investment Group, British Rail Western Region, Marks & Spencer plc, The Wellcome Foundation, British Telecom, British Airways and The Post Office over a three year period. There are five texts covering English, Science, Mathematics, Technology, and Geography, each text containing ten or twelve units or scenarios based on activities in industry, commerce or public services. The Editors/Project Directors are most grateful to both schools and industry for their cooperation in producing these resources. The materials are designed to provide teachers with background information about 'world of work' situations to which they would not normally have access.

Sponsored by:

Clerical Medical Investment Group
British Rail (Western Region)
Marks & Spencer plc
The Wellcome Foundation
British Telecom
British Airways
The Post Office

Project Directors and Editors:

	Prof Ken Austwick	University of Bath School of Education
	Dr Eon Harper	EHC
Authors:	Gerald Lloyd	University of Bath
	John Trickett	University of Bath

Concrete – the castable stone

Can you imagine a kiln, like a great submarine, but almost twice the length of a soccer pitch? Powdered and burning fuel is being forced in at one end to be heated to about fourteen times the temperature of boiling water – so hot that it would melt silver. This kiln slowly rolls over, on the spot, once a minute.

This is the giant that was under the control of Derek Arnold while making cement at the Blue Circle Works at Westbury in Wiltshire.

In a year the Westbury works can make up to 700,000 tonnes of cement. All of this must be packaged and delivered.

Derek Arnold suggests two projects. The first involves designing a suitable form of cement packaging for DIY shops. The second involves the design of concrete furniture for use outside the firm's sports pavilion.

BACKGROUND INFORMATION

Information about the cement works at Westbury is presented by the company's former Technical Manager: the nature of his job, the need for concrete, and the industrial process.

The assignments are presented to students by this Technical Manager.

STUDENTS' FIRST ASSIGNMENT AND THE NATIONAL CURRICULUM

Assignment

A factory can produce 700,000 tonnes of cement in a year. A large proportion of this is delivered for sale in quantities of approximately 50 kg to DIY establishments. A means of packaging has to be devised.

Programme of study

Propose modifications to improve the performance and appeal of existing products, such as how to make carrier bags stronger, more attractive and more resistant to rain.

Attainment Targets

Identifying needs and opportunities (1.7a). Students should 'analyse information of several kinds and draw conclusions about the needs and opportunities for a design and technological activity'. ACTIVITY: several examples of packaging are provided, and these should be analysed against what is considered important in the packaging of cement.
Generating a design proposal (2.6a). Following this analysis, students can 'produce a design specification and use it to develop their design proposal'. ACTIVITY: various materials are given that could be used for packaging. These are accepted or rejected on the basis of the constraints specified previously, and a more appropriate form of packaging is devised.
Planning and making (3.5d). Students should 'apply knowledge of materials, components and processes to overcome problems as these arise'. ACTIVITY: a mock-up is constructed.
Evaluating (4.5b). Students should 'justify the materials, components, procedures, techniques and processes used'. ACTIVITY: the factory's Board of Directors requires a report 'suitably written, and illustrated'. This should include an evaluation of the artefact. You could ask others in the class to play the Board of Directors receiving and commenting on the report.

STUDENTS' SECOND ASSIGNMENT AND THE NATIONAL CURRICULUM

Assignment

The National Curriculum's requirement that students should work with a range of materials can be achieved with the help of this unit, the material being cement/concrete. If your students have not mixed, moulded and manipulated cement then they should have an opportunity to do so before commencing this activity. You will be familiar with the caustic properties of concrete/cement mixes and understand the need for the usual safety precautions. Students are required to design furniture for a particular environment, in this case the patio area of the factory's social club.

Programme of study

- Develop and apply understanding of how materials are shaped, cast, joined and formed.
- Recognise the importance of the views of users and others affected by design proposals and consider them in taking decisions – e.g. furniture design.

Attainment Targets

Identifying needs and opportunities (1.5b). Students should 'recognise that social and environmental considerations and the preferences of users are important'. ACTIVITY: the furniture has to be designed for use outdoors on the patio of the social club of a large factory.
Generating a design proposal (2.5e). Students are asked to 'specify what they intend to do and what they will need by using simple plans'. ACTIVITY: specifications have to be seen by the social club's Committee for acceptance or rejection.
Planning and making (3.5b). Students must 'use knowledge and understanding of the properties of a range of materials'. ACTIVITY: working with concrete/cement – probably an unfamiliar material – extends students' knowledge.
Evaluating (4.6a). The next stage is to 'review the needs and opportunities originally identified and decide if they are appropriate'. ACTIVITY: the final design is reviewed against the original factors considered appropriate, such as stability, comfort, ease of moulding, in a report to the factory's Social Committee. You could ask other members of the class to play the roles of Committee members.

Topic networks
Students could be expected to devise topic networks about various social, economic, environmental and technological considerations, i.e. Attainment Target 1, level 5. An example is shown on page 3.

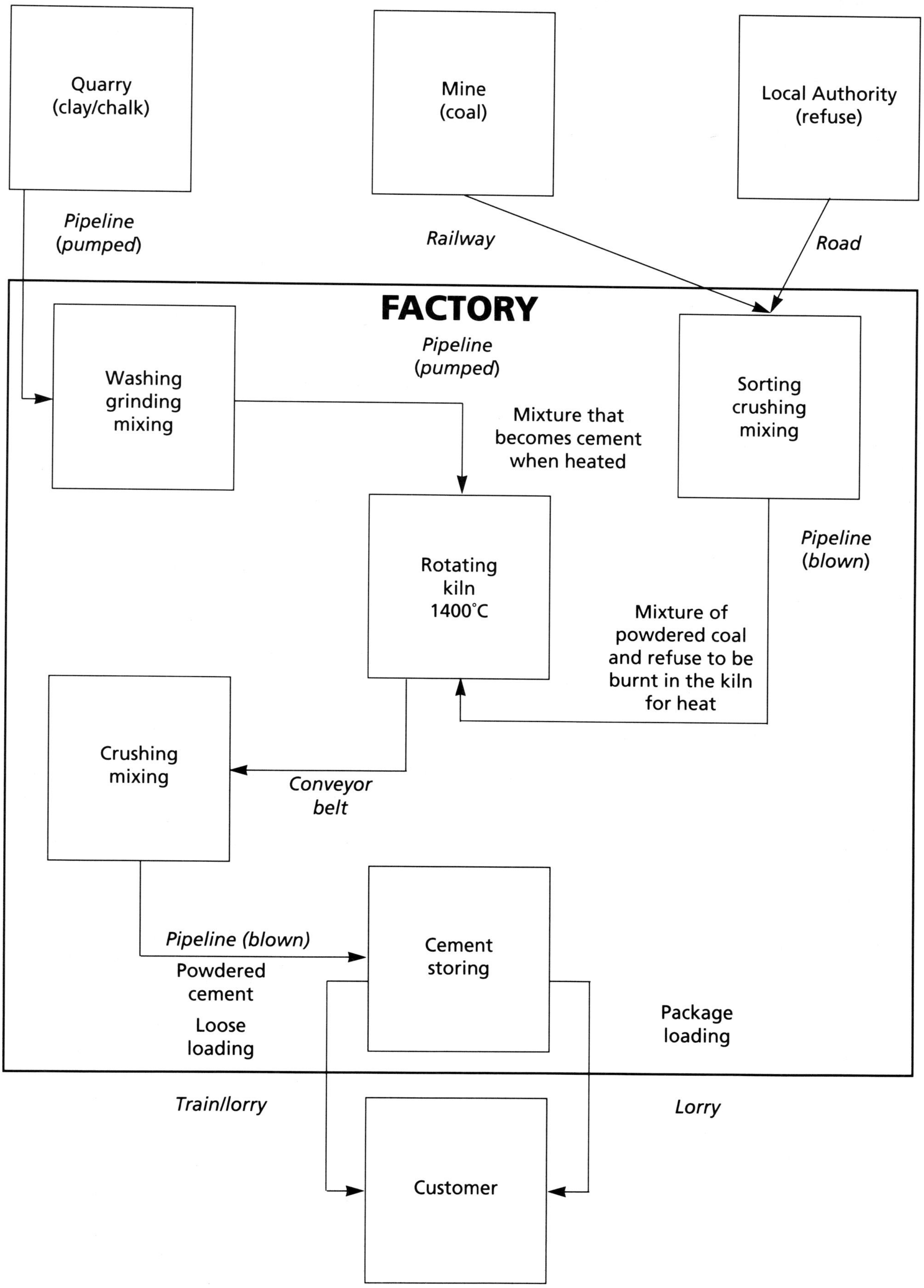

Quarry
(clay/chalk)
Mine
(coal)
Local Authority
(refuse)
Pipeline
(pumped)
Railway
Road
FACTORY
Washing
grinding
mixing
Pipeline
(pumped)
Mixture that
becomes cement
when heated
Sorting
crushing
mixing
Pipeline
(blown)
Rotating
kiln
1400°C
Mixture of
powdered coal
and refuse to be
burnt in the kiln
for heat
Crushing
mixing
Conveyor
belt
Pipeline (blown)
Powdered
cement
Cement
storing
Loose
loading
Package
loading
Train/lorry
Lorry
Customer

CONCRETE INFORMATION

The Technical Manager's job

I'm Derek Arnold. I worked at Blue Circle for twenty years, ending up as Technical Manager. Now I've retired, but the company still makes use of my skill by asking me to show visitors around the plant.

At this works there are five managers and all are responsible to the Works Manager. He has divided the tasks into groups and employs managers to see that these are done. The Technical Manager looks after the production, quality and packaging of cement. So the giant kilns were my babies!

Other managers are: a Maintenance Manager to keep all machinery working; an Engineering Manager to repair machinery; a Distribution Manager to sell and distribute our cement; and last, but by no means least, an Office Manager to organise and keep records of all that we do.

The Technical Manager now has about 40 production workers, who must all work as a team. Unless they communicate easily and efficiently with him, the Manager's direct contact with the kiln is lost. If this happens he will not be party to that strange feeling the kilnworkers sometimes have when all is not well. Of course, they may not be sure what is wrong, but that is where his skills and experience come in.

Officially, he works for eight hours a day, five days a week. However, there are people working on the plant for 24 hours every day in the year, so he has to be within easy reach of the works for most of the time and lives within a few kilometres of the plant. Unofficially, his hours are not fixed and he is trusted to put in the necessary time.

Westbury is not all work; we have a social club on-site. Outside is a patio area where members can sit in the sun. Our Social Committee is to provide new furniture for it; you might be able to design some which is attractive and, naturally, made of cement.

THE INDUSTRIAL PROCESS

The need for concrete

Concrete – a mixture of cement, sand, stones and water – will set harder than most rocks, even when under the sea. Yet, before setting, the mixture can be poured into moulds to form the simplest or the most complex of shapes.

Just think of the journey you made this morning. The building that you left is so heavy it would, without its concrete or brick foundation, sink unevenly into the ground and crack apart. Probably, its building blocks are mainly concrete, and the 'glue' holding them together was originally a mixture of cement, sand and water.

You may have travelled by road, over or under bridges, and past offices and shops, before arriving at school. As each of these structures would fall down without their cement, you can imagine that, somewhere, people must have made vast quantities – and, if we are to continue building, they must continue to make even more.

This was my job and that of my 240 colleagues working at Westbury Blue Circle. They make and sell cement, the major constituent of concrete.

You may wonder how such a factory, with its tall chimney stack, could find a place at Westbury, below the famous Wiltshire White Horse. In fact, what makes the White Horse also helps to make cement. The factory has been placed near, if not on top of, its main raw materials: chalk and clay.

Making concrete – mixing and heating

I'll take you on a conducted tour of the plant, and in return I'm going to ask you for some advice!

To make cement we heat a mixture of water, chalk and clay with small quantities of sand and iron oxide. The resulting lumpy material, called 'clinker', becomes a cement when crushed. There are many different cements but they all harden through the formation of interlocking crystals of various silicates and aluminates: these are the same chemicals that make up some of the hardest natural rocks.

We need enormous quantities of chalk and clay to produce our maximum yearly output of 700,000 tonnes of cement. This comes from the quarries on our site. After we have removed clay, the great holes that remain are filled with refuse from the local authority and covered over for agricultural use.

The heating takes place in one of our two rotary kilns. Each one is 4 metres in diameter and when cold they are abut 150 metres long. When hot they are about 0.6 metres longer, through the expansion of the metal. Fire-brick linings protect the outer steel casings from the heat.

The kilns point slightly upwards, at an angle of two degrees to the horizontal. The mixture to be heated is run into the high end and as the kiln revolves, once every minute, the mixture gradually moves down. At the lower end we blow in a burning, pulverised coal. Once we burned household rubbish with the coal: this reduced our coal costs by 10% and helped the Local Authority to destroy some of its waste. Now it is more economical to use the rubbish for land-fill.

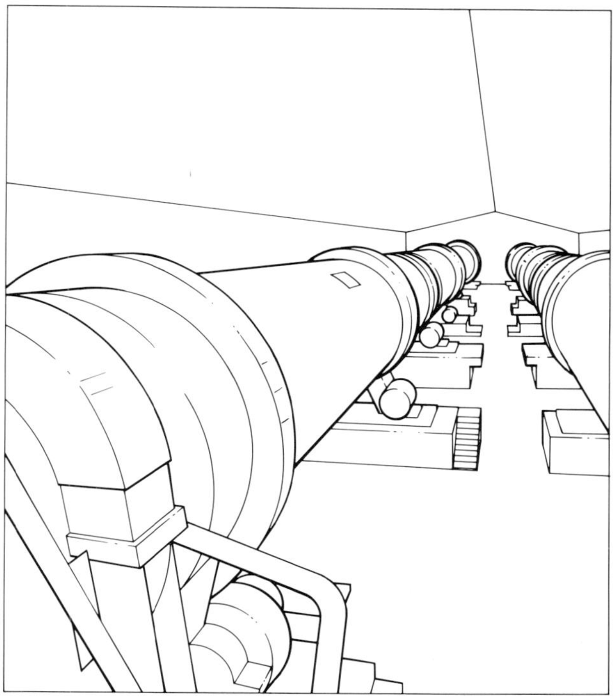

Inside the kiln hang gigantic metal chains. As the kiln turns, so the chains swing, turning and drying the hot mixture as it passes through.

Crushing, mixing and keeping down dust

At about 1400°C (degrees Celsius), the mixture is near its melting point and sticks together – we say 'sinters' – into lumps about the size of a walnut. This white-hot clinker takes about three hours to pass through the kiln. About 40 tonnes comes out of the end every hour.

The heat in the clinker costs a lot of money to produce and we cannot afford to waste it: we use this hot clinker to preheat the air we blow into the furnace.

Before crushing the clinker to make cement, we add about 6% by mass of gypsum. This slows down the hardening of the cement when it is mixed with water.

The mixture of clinker and gypsum is crushed to powder. We use about 79 tonnes of hardened steel balls in each of our four grinding mills. They are turned by electric motors with a total power equivalent to a hundred cars. When the mills turn, the balls inside roll over and over, crushing the clinker and mixing it with the gypsum.

Student Materials

Making cement creates a lot of dust. This is a major problem. The level is monitored by the Government's Alkali Inspectorate. We keep well below the limits placed upon us by using sixteen electrostatic precipitators to clean all exhaust gases from the kiln. Afterwards they are passed through our 120-metre chimney stack, a local landmark.

The cement is stored in eight 2000-tonne silos. Half (50%) is transported in bulk as a powder, 15% is taken away via a branch railway line, and the remaining 35% is taken away in 50 kg bags by lorry.

We can make 2000 tonnes of cement in a day. To do this we would use:

chalk	3260 tonnes
clay	400 tonnes
coal	400 tonnes
gypsum	120 tonnes
iron oxide	26 tonnes
sand	112 tonnes
water	2,000,000 litres
electricity	2,000,000 kilowatt/hours

The Technical Manager's first project – how should the cement be packaged?

Information – DIY

Our factory can produce 700,000 tonnes of cement in a year, of which 35% is sold in quantities that can be easily carried from a DIY shop.

We are looking for another type of packaging, designed especially so that customers can take away 50 kg bags. Remember that cement is a fine, potentially dusty, powder that must be kept dry.

The first task is to decide the factors to consider. Here are some ideas but I'm looking for others. Add to this list and then make a 'table of priorities'. Do consult other people, especially those who are likely to buy cement from your local DIY shop.

inexpensive

place for 'instructions'

biodegradable packaging

portable when full

automatically fillable

attractive

not easily burst

keeps out moisture

easily stacked/rolled before use

produces no dust when filled

recycable packaging

easily stacked when full

has appropriate place for company logo

Complete a table something like this:

Table of priorities

Factors	Order of priority
	1
	2

Look at your priorities when you have finished the table, and base your solution to the following design problem on these.

Here are some ideas for packaging. We would reject them all! Look again at your table of priorities. On what priority do you believe that each packaging has been rejected?

Jute sack – stapled strongly

Biodegradable/natural!

Plastic bag – tied with string

Recyclable/light!

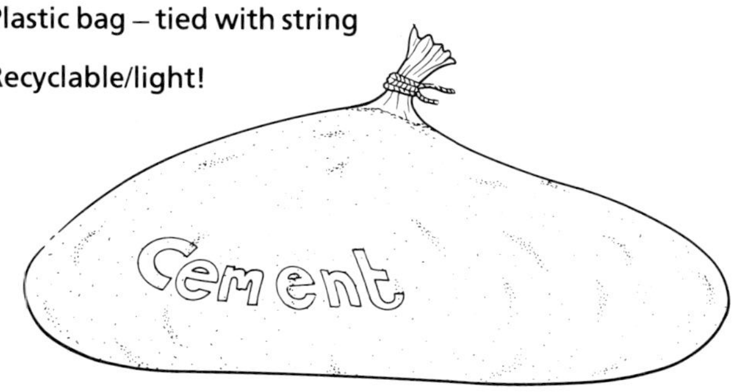

Metal drum – tightly fitting lid

Strong – keeps contents dry!

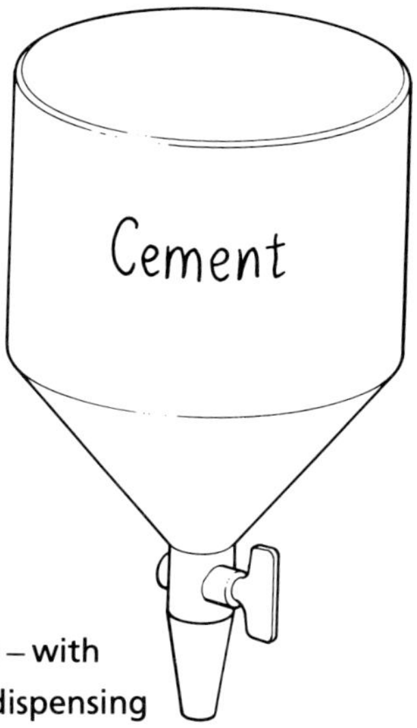

Cardboard tube – with funnel/tap for dispensing

Easily dispensed/neatly stored!

Cardboard box – packs flat

Easy to seal/stacks well!

Supplied loose – help yourself

Cheap/easily refilled!

Assignment 1 – cement packaging

Now it is your turn to come up with a better idea. Write a brief that describes your recommendations. It should be suitably written and illustrated, and include a mock-up. My Board of Directors will examine this before accepting or rejecting your designs. Of course, you should evaluate your designs against your table of priorities.

The Technical Manager's second project – furniture for the social club

You may have read how we make cement at Blue Circle in Westbury and enjoy its social facilities. Our Social Committee is considering the provision of extra furniture that can be left outside the sports pavilion in all weathers.

I need some suggested designs to present to my Social Committee for bench seats. It has been decided that they must be made of concrete. Ideally, you should have done some work with concrete and know how it can be worked. We need several designs, with drawings and a list of the advantages and disadvantages of each.

We feel there are several important factors to consider when making outdoor furniture with concrete. We have listed some, but we are sure that you will think of more, perhaps by asking your friends for their opinions.

Can the shape be moulded easily?

Will it trap water?

How can the parts be fixed together – bolts, screws, rivets, joints?

Will it be too heavy?

Are there thin parts that might shear off?

What decoration would be suitable?

Will it be comfortable?

Can parts be moulded?

Is the shape stable?

Is it attractive?

Assignment 2 – designing bench seats

We will decide the best mix of concrete as that is our expertise. What we need are several good bench seat designs to choose from, created by you. You may wish to include some scale drawings, models and an indication of why you believe these seats would be suitable for our patio area. Perhaps you could give some indication of the cost. Prepare a report and present it verbally to the Social Committee.

Airways – service sky-high

What a thought! Travelling in a Boeing 747 through the skies at a height of about eight kilometres (five miles) at speeds around 800 kilometres per hour (500 mph) to the USA; and able to choose to eat an appetiser of vegetables and lamb, a second course of fresh salmon garnished with shrimps, french beans, grilled tomato and parsley potatoes, followed by a mango and blueberry cream dessert and a cup of coffee.

How can food of this quality be served in such a way? And what an improvement on that first international passenger flight in 1919 when a single passenger was travel-sick for all of the two-hour journey from England to France, unable to think of eating or drinking!

William Cranford spends his working day making sure that the food served aboard BA aircraft is up to the expected high standard. The problems should be evident, the major one being that the dining room can be 1000 kilometres from the kitchen. He presents two assignments. The first requires the production of the menu for an economy class meal aboard a large airliner; the second involves estimating the cost of this meal.

BACKGROUND INFORMATION

The student material explains how hot and appetising meals can be served to passengers when they are many kilometres from the kitchens and travelling at perhaps 800 kilometres per hour. The first assignment – producing an economy-class menu – can be studied independently. However, the second assignment – estimating the cost of a meal – requires either that the first assignment be completed or that you provide students with a sample meal to work with. You may be asked to specify a cash limit for the meal.

STUDENTS' FIRST ASSIGNMENT AND THE NATIONAL CURRICULUM

Assignment

The Catering Unit for British Airways is responsible for supplying the food aboard their aircraft. Students are given the task of planning a suitable menu for a group of Japanese tourists, the menu card for that meal, and a 'Chef's Chat', which helps cabin staff to present the meal.

Programme of study

Recognise the needs of individuals and groups from different backgrounds, when designing for their needs; for example, the need for different food.

Attainment Targets

Identifying needs and opportunities (1.6c). Students should 'understand how the introduction of new technologies can offer new opportunities'. ACTIVITY: pre-cooking of food and the limitation on size add a different dimension to the more common problem of designing simple menus.
Generating a design proposal (2.5c) requires that students 'seek out and organise information to help them develop their ideas and refine their design proposal'. ACTIVITY: students should realise that many of the passengers catered for here are Japanese who may not appreciate European meals, and should put some effort into researching the likely preferred food.
Planning and making (3.6a). Students should 'plan and organise making in order to achieve the desired outcome'. ACTIVITY: a meal, a menu card and a Chef's Chat are all to be prepared and coordinated.
Evaluating (4.5a). Students must 'evaluate their product in relation to the design intentions'. ACTIVITY: there are several constraints upon the meal. These include the likely food preferences of Japanese tourists, that the food be light in weight and colourfully presented, and that all must fit into the aircraft's trays and, where necessary, ovens. Any evaluation must be against these criteria at least.

STUDENTS' SECOND ASSIGNMENT AND THE NATIONAL CURRICULUM

Assignment

This assignment is better treated as a continuation of the previous problem. It involves determining the cost of providing a meal to all the passengers on board a particular flight.

You may be asked to give the cost-limit for a single meal. This cost would, in practice, be added to the price of the ticket.

Programme of study

Recognise how economics affect design and technological activities and how to work to a fixed budget.

Attainment Targets

Identifying needs and opportunities (1.5b). Students should 'recognise that economic considerations are important in developing opportunities'. ACTIVITY: determining the cost of the meal designed earlier should achieve this target as ingredients that are too expensive will be precluded.
Generating a design proposal (2.5d) requires that students 'establish and check the availability of resources required, adapting their design as appropriate'. ACTIVITY: to determine costs, a visit to a local supermarket is recommended. The suggestion is made that adjustments might be required to the menu as a result of the costing.
Planning and making (3.9b). Students should then 'demonstrate how they have overcome constraints encountered in planning and making to achieve a quality product'. ACTIVITY: any adaptation of the original menu would provide evidence that this target has been achieved.
Evaluating (4.6e). Finally, students must 'illustrate the economic consequences of innovations'. ACTIVITY: an innovative but expensive meal will be unacceptable.

CATERING FOR HIGH FLIERS

We are very proud in British Airways of being the largest international airline in the world and of our reputation as 'The World's Favourite Airline'. We carry more than 17 million passengers a year to about 70 countries.

Passengers expect and receive a quality service. Besides providing a safe journey, we try to make time pass comfortably by supplying videos, music and magazines for entertainment and, of course, food and drink. Clearly, if we do not anticipate a passenger request there is no way the aircraft can stop during the flight for some extra shopping! Detailed planning is essential.

You may have travelled by aircraft, perhaps when going on holiday, and eaten an in-flight meal. If so, you will know that a flight can involve sitting in a seat for several hours. If we filled passengers with stodgy foods, they would probably begin to feel uncomfortable, so we look for a fairly light presentation of quality foods in well-balanced meals.

Like many other young people, you might prefer vegetarian food and, as long as we know beforehand, we can supply you with a special menu. If your diet requires particular foods, we will cook especially for you.

Even a Boeing 747 – a jumbo jet, 70 metres long, 19 metres high, with a wing span of 60 metres – doesn't have space on board for the food to be prepared and cooked for nearly 400 meals. Much of

the food served on board is cooked before the flight begins and for BA flights from Heathrow this will happen at the airport's Catering Unit. Nearly 80 chefs, among a total food production team of over 200, work there – preparing food for twenty-four hours each day, and for every day of the year.

William Cranford, an experienced and highly qualified chef will take you around the Catering Unit at Heathrow Airport and explain what is happening.

Planning menus

I'm William Cranford, the Manager of Food Product Development. I'm going to take you around our Heathrow Centre. No doubt you know that many chefs say people eat with their eyes. Certainly, passengers' first impression of the food they are to eat must be affected by its appearance.

As we prepare about 120,000 meals each week, to obtain our supplies from the local supermarket would immediately empty their shelves, so our buyers travel all over the world to farms and fisheries, buying huge quantities of the highest-quality food.

In a typical year we buy:
45 tonnes of salmon fillet
20 tonnes of Scotch beef
10 tonnes of strawberries
300,000 litres of milk
50 tonnes of fillet steak
30 tonnes of tomatoes
12,000 cucumbers
100,000 litres of champagne

You can see that the scale is enormous. It is greater than that of a large hotel because we cater for up to 80,000 customers a week, with all meals being eaten far away from the kitchens. Food-production planning is extremely important. We do know well in advance, however, the flight timetables that are to be repeated week after week. On long-haul routes our kitchens are prepared for four complete changes of menu per year, based on the availability of foods in season. Within each seasonal cycle, there are many menus – designed to suit class of travel and destination. On Concorde, and on European and UK domestic routes, menus are changed weekly because many of our passengers travel with us every week.

Chefs, when planning menus, search for quality, value and, where possible, something special. They will take great care in considering:

- the aircraft's destination
- season of the year
- time of departure
- length of flight
- sleeping time

Once a menu has been planned and costed, a food presentation is held with senior staff. If accepted, a dish is then photographed and a specification is written for use in the dish's preparation, for staff training, and for menu preparation and printing.

We also produce a 'Chef's Chat'. This advises cabin staff about methods of serving and oven timings, and gives tips for heating delicate dishes and preparing garnishes, as well as describing the food in an appetising way.

The Catering Unit at Heathrow

I work at Europa House, the head office of Catering Operations Worldwide, as a member of a team of culinary experts who conceive, formulate, implement and monitor airline food and menus around the world and particularly at Heathrow.

At the Heathrow Unit there are hot and cold kitchens, cold rooms, deep-freezes and larders, as well as separate preparation rooms for meat, fish and vegetables. We even have our own bakery, so our bread is as fresh as possible. These facilities are as good as, if not better than, those found in the very best of hotels, and I know because I've worked in many of them.

I could talk at length about the enormous range of dishes that we develop and prepare. Perhaps it is enough to say that almost any dish produced for a restaurant or hotel can be adapted for use in in-flight catering.

Here are some sample menus:

Economy class (summer) *Flying to USA/Canada*	*Club class (autumn)* *Flying to Africa/India*	*Economy class (winter)* *Flying to Bermuda/Rio*
Vegetable and lamb galantine	Maine lobster	Escalopes of chicken with pineapple in mayonnaise
Poached salmon with shrimps French beans Grilled tomato Parsley potatoes	Breast of duck oriental style with kumquat and ginger Broccoli Tomato Wild rice	Grilled fillet of beef with black pepper and butter Green beans Tomato Sauté potatoes
Sauté of chicken bourguignonne Peas with carrots Noodles	Grilled fillet steak with sliced mushrooms Beans Tomato Lorette potatoes	Flounder with prawns Broccoli stuffing in white wine sauce Spinach Tomato Saffron rice
Mango bavarois with blueberries/redcurrants	Poached fillet of salmon garnished with potato and chive salad, tomato and asparagus	Fruit bavarois roulade
	Apple and blackberry croustade, double cream	
	Double Gloucester Clos l'Abbé Stilton	

In case you don't know all these foods, here are brief descriptions of three of our specialities:

- Lamb galantine is a lamb meat loaf with paté on the outside and minced lamb on the inside, that has been steamed or poached.
- Lorette potatoes are mashed potatoes mixed with choux pastry, deep fried to make it light and puffy.
- Mango bavarois is mango with Bavarian cream; a form of mousse, whipped and frozen.

I'll use trout, a 'fish that flies well', as an example of our cooking methods. The majority of airline food is cooked in advance then rapidly chilled in the Catering Centre. One of our machines will cook 1200 trout in an hour. The machine uses filtered groundnut (peanut) oil and turns the fish over as they are being fried. Final preparation, using the Chef's Chat, takes place aboard the aircraft. Large aircraft have convector (air-flow) ovens that are normally used to heat food for 25 minutes at a temperature of 150°C.

The afternoon flight to Boston

As on every other day, a flight is to leave Heathrow for Boston in the USA. The Boeing 747 will take off at 15:15, arriving in Boston at 22:40 UK time (17:40 Boston time) – taking about seven hours.

On checking the bookings we find that there are 18 first-class, 90 club-class and 258 economy-class seats taken. We need to produce 366 meals which must include, or so a computer print-out tells me, two Asian meals and eight children's. In addition, there are requests for one low-cholesterol, five vegan, and one fruit-only meal.

Food requiring final cooking or reheating will be supplied in dishes and loaded onto trays which slide into portable oven racks. These racks are loaded into the on-board ovens when they arrive at the aircraft with everything else that needs to be loaded. It is very important that every hot main course contains the same overall weight of each food, for example, potatoes and two other vegetables, so that they can all receive their final cooking at the same temperature and for the same length of time. There are 16 oven racks with spaces for 25 meals in each rack. The space for each tray is only 4 cm deep, so the food cannot be piled higher than that. Other trays contain food that just needs de-chilling, or no cooking at all.

After samples are taken for on-site laboratory checks, all food is taken to a pre-assembly area together with thousands of other separate items. There they are checked before being packed onto purpose-built trolleys. The list seems endless but included on it are pre-cooked meals, trays, drinks, tableware, blankets, linen, toiletries, reading material, packs of children's games, duty-free stocks, utensils for feeding babies, first aid kits, air fresheners, magazines and newspapers.

Everything is taken to the aircraft in large loaders that rise hydraulically to a height of just over 5 metres (18 feet), the door level of a 747. Each is insured for £20,000,000 – not for the value of the vehicle itself or its cargo but to cover the cost of an aircraft should one be damaged by the loader as it passes.

At the end of the flight, all waste material needs to be unloaded and sorted. Left-over food is discarded, but crockery and cutlery is fed into seven huge washing machines. After washing, everything is sterilised and blown dry, ready for inspection and starting all over again.

A TASTE OF AIRBORNE CATERING AND ITS CHALLENGES

The Development Manager's first project – a meal for economy class

I must consider:

- departure/arrival time
- how to make the meal attractive/colourful
- change of time zones during flight
- how to use locally available food
- time of year
- travel group: economy/club/first

LON

From	To	Days 1234567	Depart	Arrive	Flight number	Aircraft /Class	Stops	Transfer Information Airport	Arrive	Depart	Flight number	Aircraft /Class
FROM	**LONDON CONTINUED**											
▶	**TOKYO**											
		1----6-	1200④	0855†	BA021	747/FJM	0					
		-2-4---	1200④	1130†	BA017	747/FJM	1					
		----5-7	1200④	1445†	BA005	747/FJM	1					
		Daily	1430④	1125†	BA007	747/FJM	0					
▶	**TORONTO**											
		Daily	1015④	1315	BA093	747/FJM	0					
▶	**TOULOUSE**											
		Daily	0900①	1140	BA376	73S/CM	0					
▶	**TUCSON**											
		Daily	0955Ⓝ	2015	BA229	D10/FJM	1	DFW	1415	1848	DL1193	73S/FY
▶	**TUNIS**											
		1--4---	1045Ⓝ	1435	GT402	737/CM	0					
▶	**TURIN**											
		Daily	1000①	1250	BA580	73S/CM	0					
		Daily	1710①	2000	BA584	73S/CM	0					
		Daily	1755②	2040	AZ1293	DC9/CY	0					
▶	**VALENCIA**											
4 Dec–30 Mar		-2--5--	0915Ⓝ	1230	GT142	737/CM	0					
▶	**VANCOUVER**											
		-----6-	1340④	1520	BA085	747/FJM	0					
		-2-4---	1340④	1720	BA085	747/FJM	1					

Illustrated is part of a page taken from an airport timetable. Your concern is with flight BA007 from Heathrow. Find out the time of departure, and the time and place of arrival.

The on-flight meal will be served in dishes shaped as shown. You can see that there isn't much room in the dishes. When the meal is planned another important factor to consider is the height of the food when in its dish. All our food is pre-cooked and chilled before it is loaded onto the aircraft to be reheated in ovens. The ovens take trays but these cannot be deeper than 4 cm when loaded. Large baked potatoes are therefore quite impractical.

All dishes, food and utensils fit into one tray.

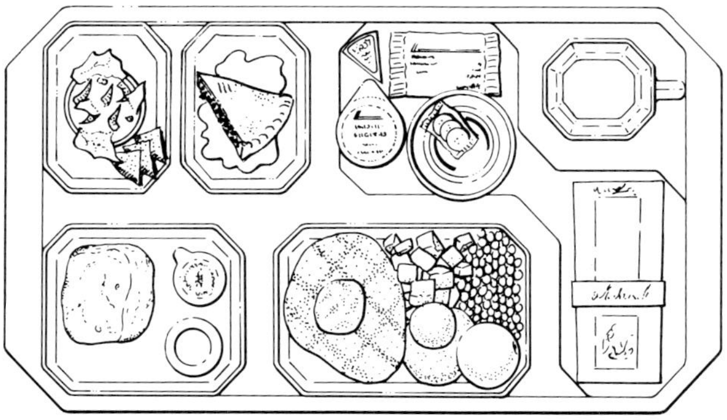

Again, there isn't much room, particularly for the main course with its two vegetables. The tray has been designed to take three dishes and has places for bread and butter, cheese and biscuits, a glass, a cup, a powdered milk sachet, and a packet which holds cutlery, salt and pepper, sugar, a serviette, a toothpick and a 'cool wipe'.

Make an intelligent guess as to the nationality of most of the passengers on flight BA007 and research into the type of food that they might prefer. Consider the places they might have visited and what they might have seen while in the UK. Now imagine the meal they would enjoy at the time of departure, while travelling to that destination after a holiday in the UK.

Assignment 1 – a meal for economy class

Plan a meal to be served on flight BA007, bearing in mind the limits to cost, weight and volume. Don't forget that an attractive and colourful meal makes a considerable contribution to most diners' enjoyment. You could evaluate your menu with this criterion and the nationality of your passengers in mind.

Devise an illustrated menu card with an appropriate theme and, to ensure that your meal is served in the manner you intend, write and illustrate a Chef's Chat that indicates exactly how the food should be cooked and arranged by cabin staff before being served. The Chat should cover no more than a side of A4 paper, and should include photographs or illustrations and reasons why the menu is suitable for those it has been devised for.

The Development Manager's second project – evaluating the costs of an in-flight meal

A meal has to be costed before it can be adopted. To make it simpler to cost your meal, imagine that for this particular aircraft there are only economy-class seats and the meal you have designed will be served to each passenger. The first task is to decide how much food has to be ordered for this one flight. Perhaps the easiest way is to multiply the weight of all the ingredients for the meal by the number of passengers. Making an estimate of the waste that is unavoidable in preparation is also necessary. The seating plan may be of help.

It is also necessary to find the cost of all the ingredients for the meal. If you do this you will be able to make a cost-evaluation of the success of your design. This might involve a trip to a local supermarket. You may decide to adapt your menu on finding that particular items are too expensive and that there are adequate and cheaper alternatives.

If you could find a way to buy in bulk this might reduce your costs. If you decide against this, you could compare the prices of two supermarkets to find which offers the better buys. Don't forget all the extras such as salt and pepper, bread, butter, coffee/tea, sugar and milk.

Of course, if you were running a small restaurant, you would have to add in many other costs. These would include those for staff salaries and uniforms; advertising and menu production; interest on a loan you might have taken to set you up in business; equipment depreciation, cookers, washers, crockery, cutlery, glasses; heating and lighting; insurance; and that's not all! I don't want you to add in these costs: our overheads are even more complex and expensive!

Assignment 2 – costing your meal
By working out the number of passengers on flight BA007, calculate how much food is needed for your menu and find out how much it will cost. Compare the total cost with the limit given by your teacher.

Note-for your assignment
assume all seats are 'economy'

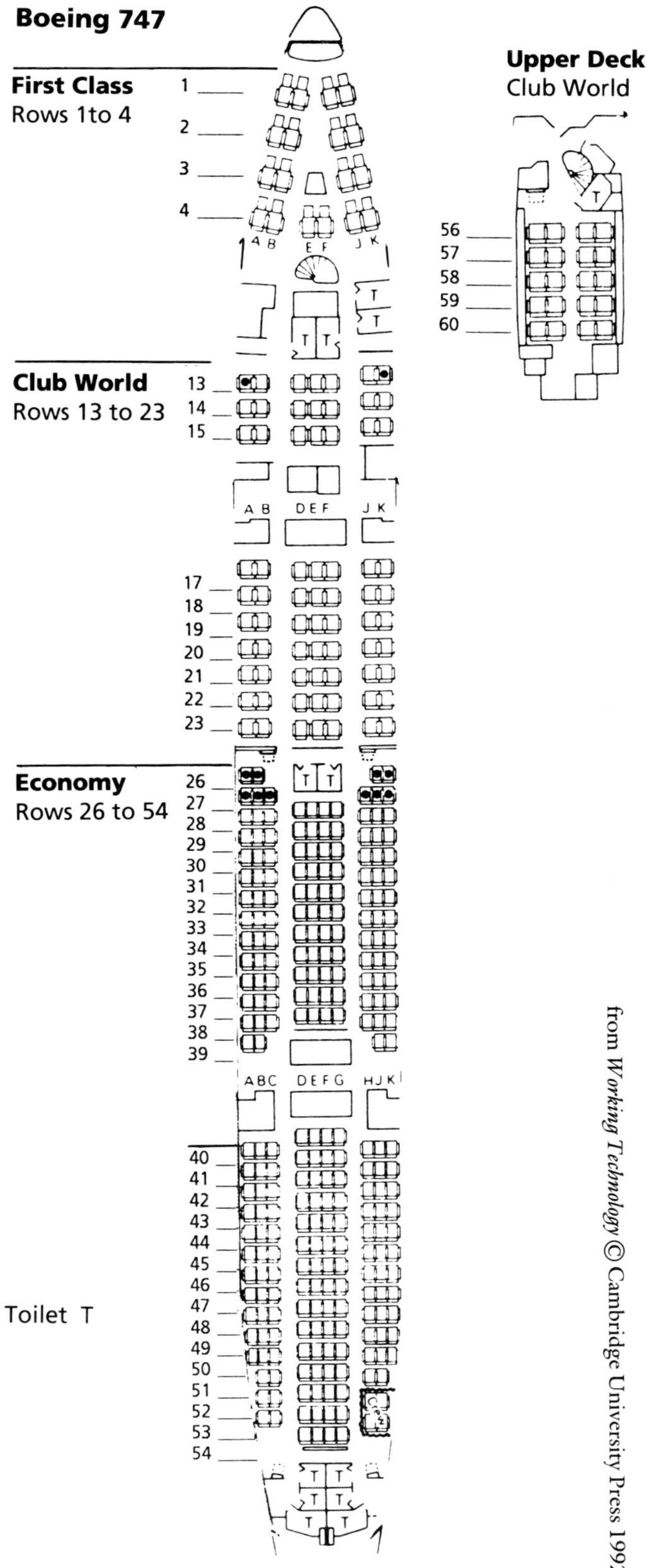

Rail technology – the way forward

A train, propelled by two engines that together have the power of about 200 small cars, travelling with 400 passengers along two parallel steel rails, must stop at a precise spot within centimetres of a long platform crowded with people at Bath Spa station.

Waiting there is the Station Supervisor, Jo Mehew. With her colleagues she has to see to all parcels and ensure that passengers board and leave this train safely, within the two minutes allowed for stopping at the station. Then, on her signal, this high-speed diesel–electric train will accelerate off to speeds of nearly 200 kilometres per hour (125 mph) in seven minutes.

Nowadays, we take for granted this amazing technological achievement as it happens time and time again throughout the country.

Jo Mehew suggests two assignments. The first involves deciding what information passengers need on arriving at a station platform and designing a screen or notice boards to contain this. The second requires a design for a new station – a very intricate task.

BACKGROUND INFORMATION

The Student Materials include sections describing advances in rail technology, the demise of steam power, the development of stations, and the everyday tasks of a Station Supervisor.

Two assignments are provided with at least some sections of the first needing to be completed before the second can be attempted.

STUDENTS' FIRST ASSIGNMENT AND THE NATIONAL CURRICULUM

Assignment

Within the context of a railway system, the problem arises of a Station Supervisor, having only two minutes to ensure the safe arrival and exit of an HST (High Speed Train), requiring some means to reduce the number of questions asked while the train is at the platform and yet ensure that members of the public have the information they require. This requires an investigation to discover what information railway passengers need, and then to display this in a way that does not exclude its use from those who have difficulty with the printed word. The activity could be a simple pen-and-paper exercise or could involve the use of more sophisticated aids to designing.

Programme of study

Use symbols and conventions that have meaning for an international audience.

Attainment Targets

Identifying needs and opportunities (1.6c). Students should 'understand how the introduction of new technologies can offer new opportunities and create new demands for design and technological activity'. ACTIVITY: the students must decide what railway passengers would find useful and design a screen/board that would take this information.
Generating a design proposal (2.8b). Students are required to 'plan their activities to take into account multiple constraints which may at times be conflicting'. ACTIVITY: there is much information that could be put on the screen/board but it is limited by size; and different passengers have very different demands, as the students may realise by comparing lists.
Planning and making (3.8a). Students should 'review how to make best use of materials, procedures, tools and equipment'. ACTIVITY: the classroom limitation must be considered. Computer assistance in design work may be available in some classrooms while not in others.
Evaluating (4.8a) requires the students to gauge 'the suitability of the product for manufacture'. ACTIVITY: the Station Supervisor asks for a report as to whether the final product could be manufactured on a large scale. The result would be passed to the British Railways Board. You might take on this role in accepting and commenting on the report.

STUDENTS' SECOND ASSIGNMENT AND THE NATIONAL CURRICULUM

Assignment

A railway station must be designed that satisfies given criteria and a plan to scale of the design must be produced.

Programme of study

Analyse environments by relating knowledge of design history to students' own observation and interpretation; for example, consider designs of Victorian railway stations when presenting plans for the renovation of a local station.

Attainment Targets

Identifying needs and opportunities (1.6b). Students should 'explain how different cultures have influenced design and technology, both in the needs met and opportunities identified'. ACTIVITY: given the design of a simple Victorian station, a student should gain some ideas about designing a modern station and how demands have changed.
Generating a design proposal (2.6d) requires students to 'use specialist modelling techniques to develop design proposals'. ACTIVITY: the techniques used must depend upon the facilities available within the classroom and the abilities of the student – but the potential exists.
Planning and making (3.7c). Students should 'use a range of technical, symbolic and other means of representation to assist in planning, organising, making and incorporating necessary modifications'. ACTIVITY: in this case, the modification is to a railway station.
Evaluating (4.5c). Students must 'understand that innovations from other times and cultures have identifiable characteristics and styles'. ACTIVITY: a comparison can be made with one of Brunel's Victorian stations.

RAIL TECHNOLOGY – PAST AND PRESENT

Moving very heavy objects by rolling them on tree trunks is a simple but clever idea. No one knows who thought of improving the design by cutting off sections of the log and joining them in pairs with an axle to make a pair of wheels.

The force of friction that must be overcome when moving heavy objects, even on rollers, is great. It has been reduced by using horses to pull loaded barges over the nearly frictionless surface of water in canals, but this is slow. In the eighteenth century, producing a road with surfaces like those of today was not technically possible. A smooth, though not too smooth, surface, and a faster mechanical 'horse' had to be invented!

A narrow path of metal, one for each line of wheels, was a possibility. Even so, some way of preventing the wheels from leaving the rails was essential. The Babylonians, as far back as 2240 BC, had had the idea of building parallel paths of smooth stones. These had grooves to prevent the wheels coming off. Later came wooden rails and wooden wheels with a raised edge, called a flange. However, the wood soon wore out. Advancing technology provided new and cheaper metals. Metal wheels could be moulded with a flange and lasted longer. You probably know what prevents a modern railway engine's metal wheels from leaving the track.

In about 1850, steel rails replaced those of cast and wrought iron. The success of this idea can be judged by considering the millions of kilometres of steel line that are used nowadays. The longest continuously welded rails are in Australia. They are about 40 kilometres (25 miles) long. To prevent them buckling through expansion in a summer temperature of 40°C, they are stretched with a tension of 120 tonnes before laying. What a technological advance on those Babylonian grooved stone blocks!

Perhaps the most original advance came with the steam engine. In 1800, they were used mainly to pump water from mines. It was Richard Trevithick who had the inspired idea of using this power to turn the wheels on which the engine itself was placed. So appeared the first steam locomotive. Its successors were to replace horses as the best way to haul heavy loads and become the basis of our original railway system.

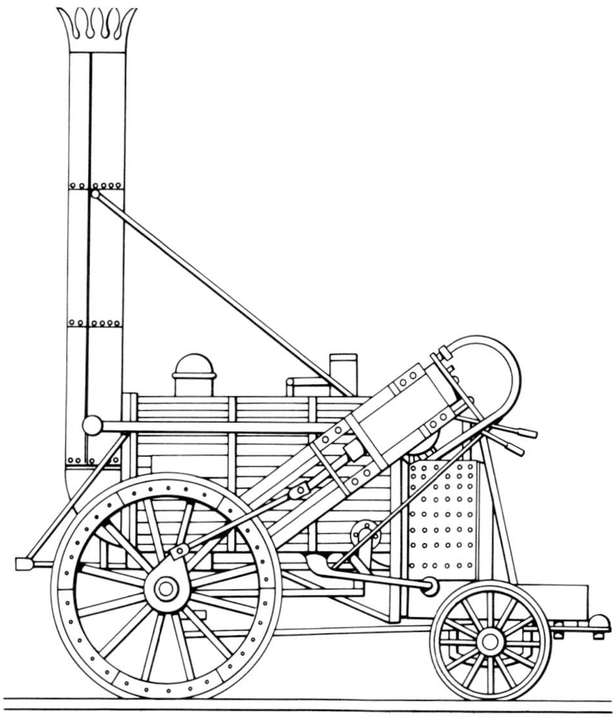

Competing technologies – steam/diesel/electric

Originally, locomotives' wheels were turned by the energy that is released when water boils to steam. The heat needed can come from burning coal or diesel oil, though dried fish, llama dung and dried moss have also been used.

In the late eighteenth century, cheap coal provided the heat that Richard Trevithick experimented with to power a steam engine. In 1808, his fourth locomotive, *Catch Me Who Can*, was used in London's Euston Square to show the potential of railways for passenger transport. Unfortunately, the track was then just not good enough.

In those days, there were fears of pollution, danger to people, and that 'the metal wheels will never grip the rails'! Over the next 150 years, the technological advances in the design of steam locomotives were gigantic.

Many people still find steam-powered locomotives highly attractive when compared with diesel and electric power but that romanticism cannot overcome what others believe are a steam engine's disadvantages of lower efficiency, poorer acceleration, higher pollution, and lower reliability.

Great efficiency results when diesel oil is used to produce electrical energy. Locomotives using this principle are diesel–electrics. They are powered by a motor driven by electricity produced from an on-board generator fuelled by diesel.

Even so, the most efficient way of producing electricity is in large power stations. This leads naturally to trains that run on electricity taken from the national grid system. Evidently, the electricity has to pass from the power station to the moving locomotive. Some engines pick this up from an electrified line that runs down the centre of the track. Others employ a system where the electricity is picked up from an overhead line.

At present, technology is advancing at an enormous pace. Future means of transportation could include overhead monorails and hovertrains, as well as the use of gas turbines and linear induction motors. Whatever the advances, railways with double-track high-speed trains can carry more traffic than a six-lane motorway and, as these systems occupy only a third of the land used by a motorway, they offer an attractive way forward.

Places to stop

Evidently, if a train is to carry passengers and freight, there must be stations so that loading and unloading can take place.

In Victorian times there were many railway companies, mostly working in competition. As even the gauge of the lines themselves could vary from one company to another, it is not surprising that stations came in a variety of shapes and sizes.

Many stations were built and many, though not by any means all, were grand and imposing. They were intended to impress both passengers and potential investors. In later Victorian times, the spread of rail was such that previously remote regions were reached. Unstaffed stations and halts began to appear, with some being little more than a platform, if that!

The size of a station depended to some extent on the importance of the town or village it served. The basics consisted of a booking office, a waiting room, and an office for the ticket clerk. The clerk dealt with everything, including both passengers and freight. In one afternoon this could include selling a first-class ticket to a local landowner, despatching a crate of live chickens to a local hotel, checking that the oil lamps were full and lit in preparation for the arrival of the night express from London and, most importantly, ensuring that the express passed through safely.

The greatest railway engineer is thought by many to have been Isambard Kingdom Brunel. At 28 years of age he was given the task of designing and building the main line from London to Bristol for the Great Western Railway (GWR). He also designed stations; Paddington station, opened in 1854, is based on his original design. Brunel produced designs for smaller stations too, similar to the one shown here.

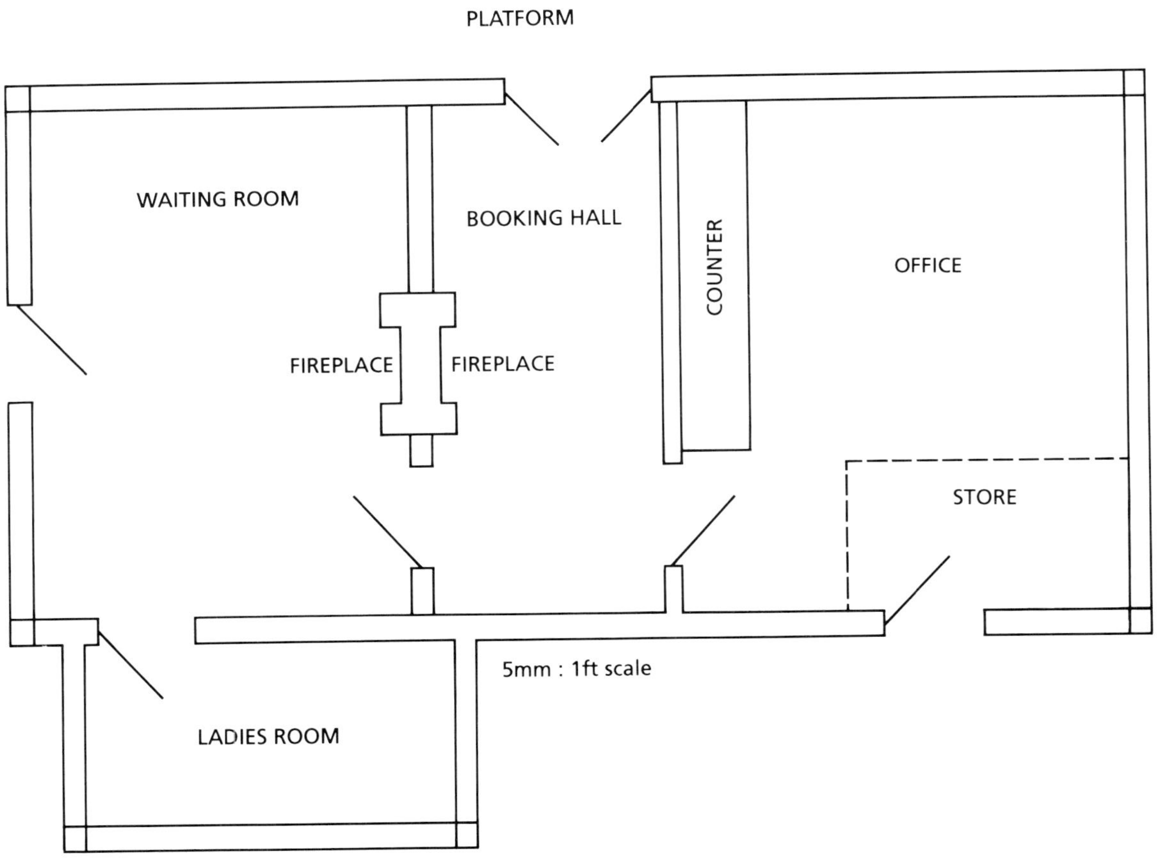

A WORKING DAY FOR JO MEHEW – STATION SUPERVISOR

Jo Mehew is responsible for the safety of the trains at her station. She's a shift worker – one shift is from 06:00 to 14:00 and another is from 14:00 to 22:00. Two people work for her, one for each of the two platforms.

Imagine you are with Jo awaiting the next train to arrive. It is a 125 to London. You are in her office as she receives information on a VDU about the next train. Much of this she types into a computer for display on platform screens for passengers. Of course, some may be unable to read the notice because of short-sightedness or not being able to read English, so she announces over the loudspeaker system the platform the train is to arrive at, where it is going and where it will stop on its journey to London. She now adds some more information: where the buffet section is positioned and, as she can see on another screen the now empty waiting room, that some cases have been left behind.

Jo's assistant is on the platform. He positions himself where he knows the end of the train will stop. The train thunders into the station. While he helps to load or unload parcels, some passengers join or leave the train. He walks down the train, checking that all is well and that the doors are safe, and signals to the conductor on the train when it is OK to depart. The conductor buzzes the driver, and away the train goes. All has been safely completed in two minutes!

How would you cope with some of these problems?

Message received beforehand from Central Control
-disabled person arriving on this train-
-Coach C, Seat 02D-
-needs BR wheelchair and ramp to leave train

Is this the 9.15 to Paddington?

Don't let the train go. I've left my brolly on board!

Wait! Wait! Open the door!

Where is the station lavatory?

I've missed the train! I'm sure it went early!

How do I get to platform 2?

I hope the train won't leave late!

The Station Supervisor's first project – information for travellers

One of my tasks is to be sure that passengers join and leave their correct train safely. As I am given just two minutes' stopping time for each train, I want to have to answer as few passengers' questions as possible and yet be sure that all have the information they need.

We have visual display units for conveying information to passengers. The screens, which can be updated almost instantly, have come to replace the notice boards we once used. The design may look impressive but I wonder whether you could improve on it in any way.

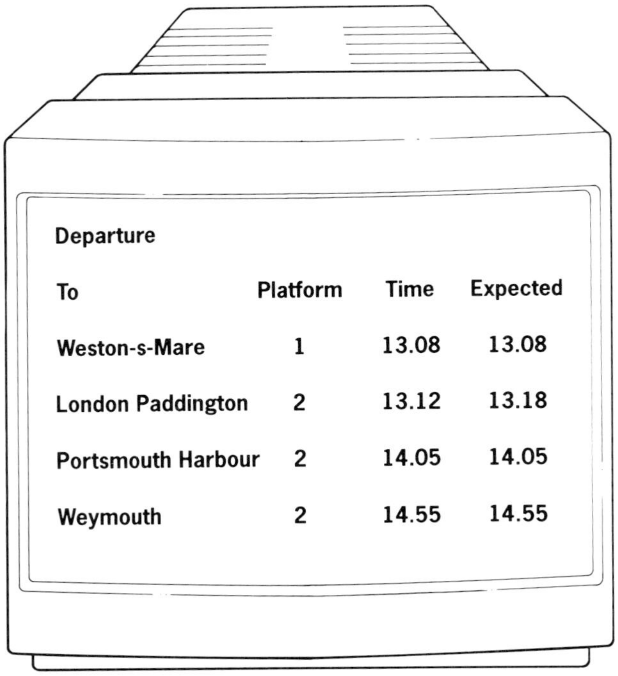

Assignment 1 – information for travellers
Imagine you are a passenger entering Bath Spa station and wanting to catch a train to London Paddington. You know the train is to leave in the next fifteen minutes or so. Make a list of what you would want to pick out at a glance from a screen or notice board. Now design a screen or notice board. Here are some factors you should consider:

- the general design – size, shape, colour, content
- that some people have difficulty in reading
- both old and young people can be very short-sighted
- positioning of screens/boards on the platforms
- that travellers from overseas may find our language difficult
- whether the design should include colour

Before you finalise your design:

- evaluate your system against your original list
- compare your design with your local station's

Present your plans to the Board of British Rail with details including a sample layout, the results of your evaluation, and an indication of whether it could be produced on a large scale. This is a chance to present your ideas with style.

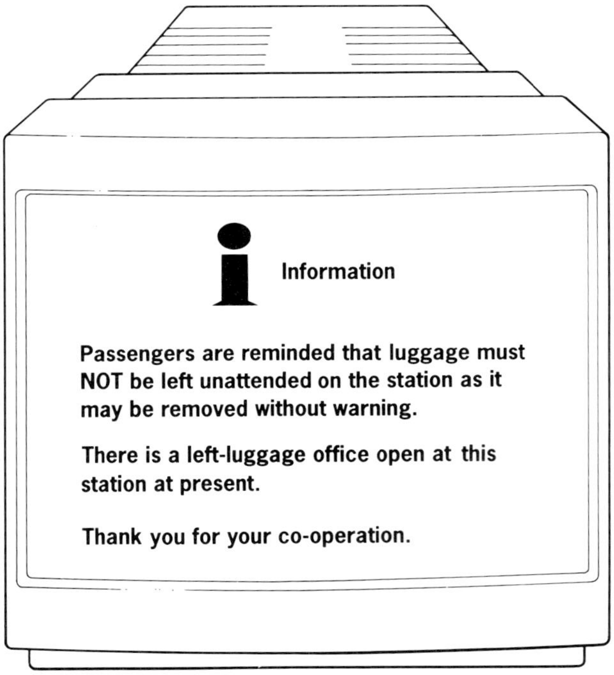

The Station Supervisor's second project – designing new stations

Our passengers do more than just get on and off trains. As a result, although small stations are quite simple, large stations that serve towns or cities offer quite a wide range of facilities.

What do you think the railway station should provide you, the traveller? You will need to make a list before you can attempt this project.

To give you a start, here are some of the questions I am asked.

Seeking information
Where can I buy tickets?
When does my train leave?
Where does my train leave from?
How do I get to platform 2?
Will there be someone to help me?

Buying tickets
Which is the best way to travel by train?
Which is the cheapest way to travel by train?
Can I eat on the train and is there a sleeper?

Waiting
Where can I wait?
Is it warm?
Is there a lavatory?
Will I have a seat?
Is it dry?
What is the time?
Can I buy something to read?
Will there be anything to do?
Will the train go without me?
Where can I safely leave my luggage?

Boarding the train
Is this my train?
Will it be late?

Leaving the train
Am I at the right station?
Will I get off in time?
What about my disabled companion?
Will it be light enough to see?
Is the bus station nearby?

Refreshments
Can I buy food?
Can I buy a drink?
Should alchohol be on sale?
Should cigarettes be on sale?

However, if you used only the above information to design a station you might have problems – for example, what if the platforms were too short to take the trains? There are also other factors to consider when designing stations: think carefully when making your list.

Assignment 2 – designing a new station
Design and draw to scale a plan of a station suitable to replace the one nearest to where you live.

Assume that the fastest, heaviest and longest train using your station in the next ten years will be an Inter-City High Speed.

Also assume that there will be two lines and that each day 90 trains will stop at the station, with on average 50 passengers joining and 50 leaving each train.

Will you need platforms? If so, how long and how high will they need to be, and how will passengers move from one to another – a bridge, a tunnel? What about those with heavy luggage or disabilities?

An important consideration is keeping costs as low as possible. Nevertheless, if your station is not attractive to your passengers, they might travel by other means: if your station is not convenient for staff, they might soon leave and get work elsewhere.

Do not attempt to estimate costs in your plan, but do keep them in mind. If your design is obviously too expensive it will be rejected by the Railways Board.

Read again the list you made earlier. How large an area will you provide for each facility? How will these facilities be arranged around the boarding area?

Finally, make an evaluation of your design and suggest how it might be further improved. This can be completed partly by visiting your local station and taking a careful look around. See how its design has overcome some of the problems you have attempted to solve. Have you done better?

Telephones – communicating worldwide

In the 1870s, Alexander Graham Bell found a way of using electricity to send the sound of his voice down a pair of thick copper wires to a room a few metres away. British Telecom is today able to send the sound of 8000 different voices, using pulses of light, down a pair of thin glass fibres from the UK to a room on the other side of the Atlantic Ocean. Not only can BT send voices but they can also send data from a computer, printed words, pictures and music.

Jilly Butler is a Service Engineer. She is very concerned with the design of telephones and call boxes as it is her responsibility to maintain them. She explains the importance of good design and presents two assignments. The first concerns a school and its payphone. The potential users and position have to be decided, a mock-up constructed and an evaluation made. The second, associated though possibly separate, assignment involves the design of a handset that could be incorporated into this payphone.

BACKGROUND INFORMATION

A very simplified idea of how sounds can be sent down copper wire and fibreglass is given, relating this to the development of the payphone system in the UK.

STUDENTS' FIRST ASSIGNMENT AND THE NATIONAL CURRICULUM

Assignment

Students will be able to read about the development of the telephone booth over the past century, finding out how the different demands of society have been significant in affecting the design of the booth. This should give students a basis from which to devise a payphone suitable for a position they have chosen within the school.

Programme of study

Recognise the importance of the views of users and others affected by design proposals and take them into account in taking design decisions.

Attainment Targets

Identifying needs and opportunities (1.8c). Students should 'investigate how needs and opportunities have led to design and technological activity in other cultures'. ACTIVITY: before payphone design can begin, students are expected to note how the designs of past booths have been affected by the requirements of the contemporary society. **Generating a design proposal** (2.9a). Students are required to 'develop ideas by drawing on information and understanding from a broad knowledge of sources and showing judgement about the detail required. Example: What problems are still likely to exist and how could the design be changed to overcome these?' ACTIVITY: Students learn about the problems that required the designing of new booths in the past. Present-day problems are examined in the context of positioning and designing a booth suitable for the students' own school. **Planning and making** (3.8c) requires students to 'identify and incorporate modifications during making'. ACTIVITY: on producing a specification, students are asked to show it to potential users, collect their comments and decide whether or not to act upon this advice. **Evaluating** (4.9a). Students must 'demonstrate that they have applied knowledge and understanding derived from evaluations of their own and others' design and technological activities'. ACTIVITY: a verbal report to the class and teacher is required that illustrates how the new design is an improvement on previous designs.

STUDENTS' SECOND ASSIGNMENT AND THE NATIONAL CURRICULUM

Assignment

Having sited and designed a booth for a public telephone at their school, students are then asked to design a suitable handset for this telephone.

Programme of study

Know that efficient mechanisms depend on the appropriate choice of materials used and the number, form and arrangement of their component parts.

Attainment Targets

Identifying needs and opportunities (1.7a) requires students to 'analyse information of several kinds and draw conclusions about the needs and opportunities for a design and technological activity'. ACTIVITY: the design of telephones through the ages is presented so that an impression can be gained of how these have changed from the simply functional to the aesthetically pleasing. **Generating a design proposal** (2.8c). Students should 'show a willingness to experiment and take risks'. ACTIVITY: the possibility is envisaged that future telephones might include live, visual images. Students taking this or some other, more experimental idea into the design of the telephone may indicate a 'willingness to take risks'! **Planning and making** (3.6b). Students are required to 'combine knowledge of a range of materials and of processes to identify those most suitable for their design'. ACTIVITY: the materials of which the telephone's main frame is to be constructed has to be specified. Students are told that simply stating 'plastics' is not enough. **Evaluating** (4.5c). Students have to 'understand that artefacts, systems or environments from other times and cultures have identifiable characteristics and styles, and draw upon this knowledge in design and technological activities'. ACTIVITY: the background to the complete project is seeing how certain artefacts, namely a telephone booth and a telephone, have changed with time. The extension of this knowledge is to define present requirements and design accordingly.

THE TECHNICAL PROCESS

Sound is a vibration which can travel through the air. We 'hear' the vibrations by means of sensitive skins, membranes, that are in our ears.

Sound can make the membrane in a telephone vibrate and this can be used to change the electric current passing along a wire. At the other end of the wire, the changes in electric current can make another membrane vibrate, passing the sound on through the air to our ears.

Nowadays, the changes in electric current caused by sound waves can be analysed and displayed by an oscilloscope. They appear as a sound wave. The wave can be represented by a series of binary numbers, each made of only two digits, 1 (one) and 0 (zero). The numbers can be passed down a wire, one or zero being the equivalent of current on or current off. The numbers can be decoded back into vibrations and then sound. Modern telephones work on this principle. They have been 'digitalised'.

Light can be used to transmit binary numbers, using light on and off to represent one and zero. Engineers have found that thin fibres of ultra-pure glass, called optical fibre, will transmit pulses of a specially pure form of light called 'laser light'. This is another way of transmitting sounds for long distances.

However, it is not always practical to have wires or fibres transmitting conversations; think of ships and aircraft! For this there are radio waves. Unfortunately, as they travel only in straight lines, to communicate over long distances they need to be reflected off something above the earth. For this, space satellites can be used.

Telstar, in 1962, was the first satellite to reflect public telephone calls back to earth. A telephone call to Australia bounces off satellites on its way, as do the television pictures from the Olympic Games and other international events that are transmitted round the world.

THE DESIGN OF TELEPHONE BOXES

You are never far away from a public phone in Britain. British Telecom operates about 85,500. All are now fitted with microprocessor-controlled push button equipment which can also diagnose and automatically report faults to the local repair centre, like the one where I am based.

In 1886 things were less organised. There were no restrictions on setting up a phone box and collecting money for the service – the British Government allowed any telephone company to build public call boxes to hold the 'new-fangled' telephone. To begin with they were treated as something of a joke. How different society was then! Someone said, 'If you really need to send a message, get a servant to carry it for you.'

Phones were placed in little wooden boxes inside hotels, railway stations and especially shops. By 1907 one of the biggest operators was the National Telephone Company. Its sign was coloured blue and

designed to catch your eye. Then, they had 7800 of their phone boxes in towns. The telephone boxes were usually made of wood and called 'silence cabinets', though silent they couldn't have been!

In those times, people were nervous of telephones. Callers had to be connected by operators and could not dial directly. Also, operators would not make a connection until the correct number of coins had been put in.

There were problems. In shops people felt obliged to buy things before using the telephone. Some shopkeepers allowed only their best customers to use the phone, and, worst of all, the shops closed at night. Free-standing wooden kiosks were therefore made that were quite independent and could be open at all times. There were two types, which came in lots of shapes. The first could not be entered unless you put a penny into the door mechanism. The second had an attendant, who took your money, made your connection and then waited outside, even in the rain, while you made your call.

As time went by, people began to see the advantage of a telephone and more and more appeared, with people wanting one of their own. The number of telephones multiplied and even more public boxes were required.

The Post Office took responsibility for the nation's telephones in 1912. They decided that red was a much more eye-catching colour and their logo was designed accordingly. The shape of boxes still varied and those on the seafront at Eastbourne were thatched!

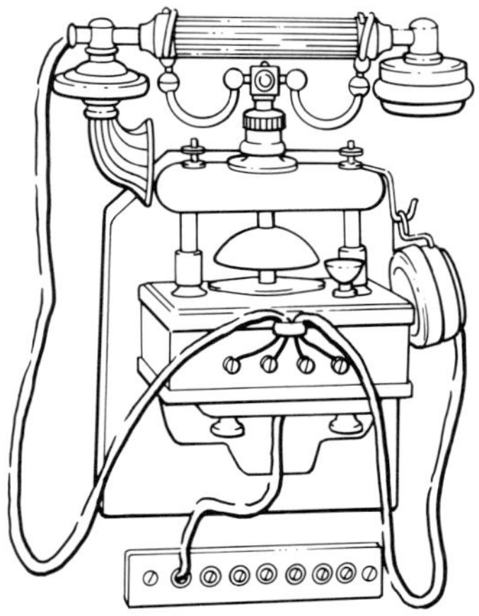

In 1921 the Post Office began its search for a uniform 'all-Britain' telephone box, and Kiosk One (K1) appeared. Made of reinforced concrete, it was pre-fabricated so that it could be put together on the spot. Unfortunately, the London boroughs did not like it and were not prepared to have it on their streets.

A competition was organised with the winner's design to be used countrywide. Costing £35 the winner was monumental, standing 2.8 metres, and being made of cast iron it weighed 1.25 tonnes. It had elegant mouldings, a flat dome and a Royal Crown that doubled as a ventilation grill. Though it was designed to be silver, the Post Office insisted upon red. This design, the K2, has since been described as a 'magnificent piece of street architecture'. Up to recent times the K2 image has remained constant but there have been modifications. The K3 was made of grey concrete but the concrete cracked, and so only 12,000 were made.

The K4, a giant, included a stamp-vending machine and a postbox. It was short-lived as the postbox made the call box noisy and the stamps became damp and stuck together. Next came the K5: there aren't even any pictures of this experimental version.

In 1935, to celebrate King George V's jubilee year, an improved design came into existence. It remained standing for many years and has been described as 'near perfection'. This, the K6, was to appear throughout the UK. It was elegant, stood nearly 2.5 metres high and weighed 0.65 tonnes. A dome remained but the windows were changed and the embossed crown no longer doubled as a ventilator. The internal fittings were bakelite and chrome. There was a mirror, a shelf, a light and even an umbrella hook. For 30 years the K6 stood out as part of the British scene.

Eventually, the time came for its replacement. Its 54 panes of glass were difficult to clean, the cast iron door was difficult to open and the ventilation was poor. So in 1960 came K7, standing at 2.2 metres, and containing the new trunk dialling payphone equipment and floor-to-ceiling windows. The cast iron was replaced by aluminium and this caused K7's failure as the aluminium was unable to stand the British weather – it turned black and blistered.

K8 was designed for new towns. Made of cast iron, it was easy to clean and paint. There were 11,000 of them put up around the country, still only a small fraction of the total of 77,000 kiosks that then existed. However, a pull of 5.2 kg was needed to open the door: a lot of effort for a disabled or elderly person.

British Telecom became an independent company in 1984. Standardisation of kiosks was replaced by diversification: now there are payphones of many different forms, each devised to suit particular circumstances – a totally different concept.

PROBLEMS FOR SERVICE ENGINEERS AND PROJECTS FOR YOU

My name is Jilly Butler, and I work as a Service Engineer for British Telecom. My job is to service and maintain telephone boxes. This task is made much easier by well-designed handsets and call boxes. The way both have changed over the years is interesting from the designer's point of view. People with my job have in the past had some difficult problems to deal with. I hope that you will help me with some suitable designs.

The Service Engineer's first project – designing an appropriate booth for a payphone

These are the features of past designs of payphone:

Silence cabinet

- 1900
- wood
- brown

K1

- 1921
- concrete
- red and white

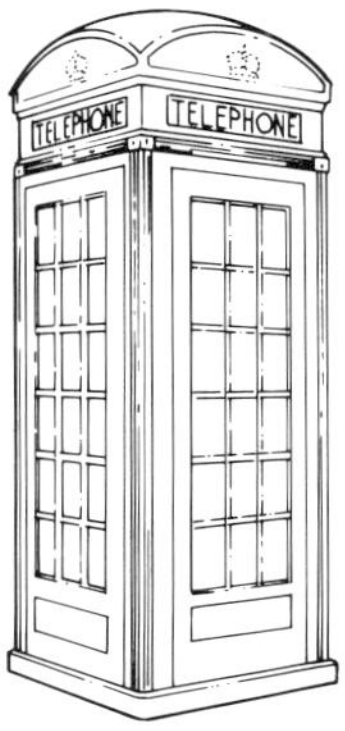

K2

- 1927
- cast iron
- red
- 1.25 tonnes
- a success

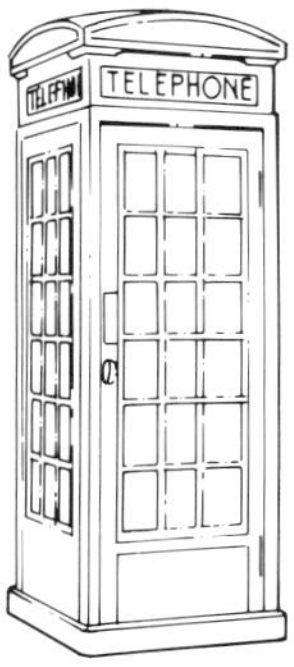

K3

- 1929
- concrete
- stone grey
- a failure

K4

- 1930
- cast iron
- red
- stamp machine and letter box
- only 59 made

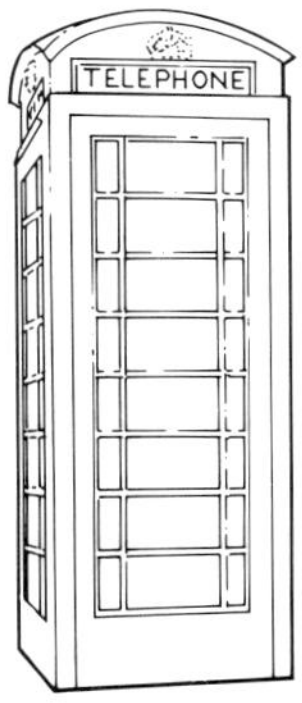

K6

- 1936
- cast iron
- red
- writing shelf
- place for parcels
- electric light
- milestone in 'good taste'

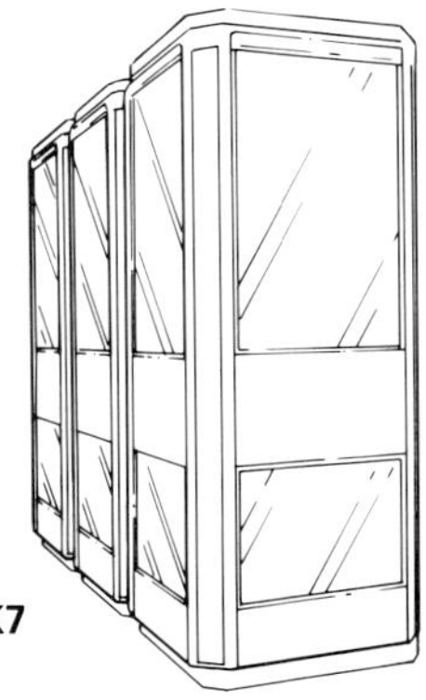

K7

- 1962
- experimental
- aluminium (only six made)
- floor-to-ceiling windows
- a failure

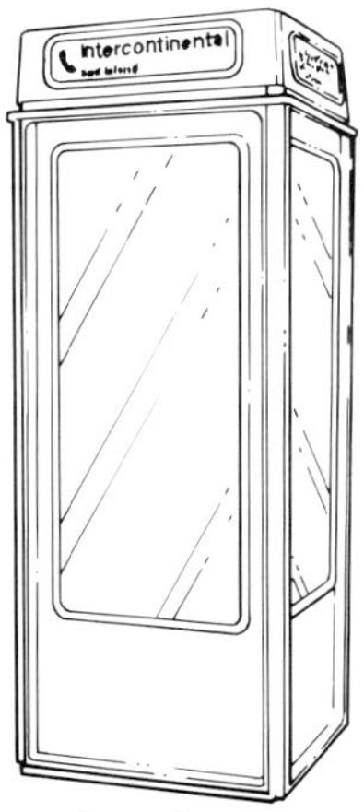

K8

- 1968
- reinforced steel
- red
- coin boxes reinforced steel
- shatterproof glass
- especially easy to clean
- a success

Nowadays, designing for particular circumstances has taken over from the standardised booth. These are some of the designs we have used recently.
We considered:

- convenience of use
- ease of maintenance
- ease of cleaning
- stopping vandalism

Assignment 1 – designing telephone payphones
This is your opportunity to design and evaluate the booth for a particular phone. Imagine that your school has been asked to design its own payphone. Decide upon an appropriate position, and devise a specification for the booth, considering both past and present designs.

Show your specification to those who might use the payphone and ask their opinion of its value. Decide whether you should change your specification in the light of these comments.

Produce your design, and then construct a coloured, cardboard model to scale. Evaluate this by allowing others in your school to inspect and comment upon the quality of your efforts.

Present a verbal account of your evaluation to your teacher and the rest of your group, describing how your design has developed.

The Service Engineer's second project – designing a telephone for a particular situation

You have probably already decided the best site for a payphone in your school and designed an appropriate booth. Another problem is to consider the design of a handset that would be appropriate for a school payphone.

Illustrated are some of the designs that have been used in the UK.

When drawing up a specification you should consider users. Who will make use of the telephone? Youngsters? Handicapped people? Short people? People with small hands? You should also consider how your design will address the problem of potential vandalism.

Assignment 2 – a suitable handset

Produce a specification, a design, a mock-up and an evaluation of your design.

The model you make could be of plasticine, but you will need to make a recommendation about the material the body of the set is to be manufactured from. Just saying 'plastics' is certainly not enough.

You might be prepared to take risks and design equipment with innovations such as live pictures of users!

Put your design on display, with an evaluation sheet to be completed by viewers. Finally, write a report about your assignment and give it to your teacher.

1924 Mouthpiece and earpiece separated; dial for connecting directly

1930 Plastics give shape variety

1950 Back to the wall

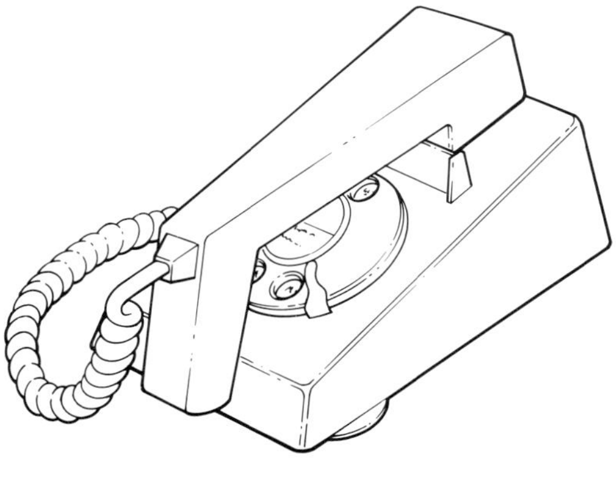

1980s No longer a 'standard' telephone

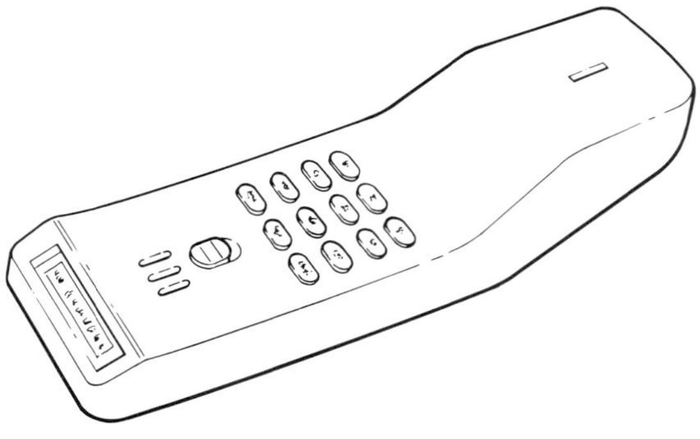

1990s One-piece and mobile

Brewing – biotechnology in practice

Although there aren't many professional brewers working in the UK, they produce about 5000 million litres (about 8000 million pints) of beer each year. The competition is fierce and a company can go out of business unless its brewers can maintain quality and devise new beers.

Philip Brunt is a Brewer. He works for Gibbs Mew at the Anchor Brewery in Salisbury as a Biotechnologist.

Here he describes his work and outlines two of the problems he encounters, which form the basis of two assignments. One of these involves setting up an advertising campaign for a new brew; the second requires the design of a sampling device for use in monitoring the fermentation process.

BACKGROUND INFORMATION

The Student Information includes a description of the industrial process at Gibbs Mew, gives the reasons for close control of the product's strength and quality, and describes the everyday tasks of a Brewer.

STUDENTS' FIRST ASSIGNMENT AND THE NATIONAL CURRICULUM

Assignment

A brewery is to produce a new beer and needs general advice about an advertising campaign, e.g. using beer mats.

Programme of study

Develop a product and a method of marketing, promoting and selling it.

Attainment Targets

Identifying needs and opportunities (1.7c). Students should 'consider both the user and the producer when defining the need for a technological activity'. ACTIVITY: a beer mat has to be designed to bring to the notice of potential customers the qualities of the new product.
Generating a design proposal (2.7b). Students then 'review the detail of their design, using their own experience and that of others, and suggest alternative ways of achieving what is intended'. ACTIVITY: the requirement is to discover what other methods of advertising are available and make a recommendation about the most suitable for this particular product.
Planning and making (3.6b). The next stage is to 'combine knowledge of the properties of a range of materials . . . to identify those most suitable for their design'. ACTIVITY: the material best suited for manufacture has to be related to the cost, size, shape and thickness of, and to the illustrations on, the final artefact, i.e. a beer mat.
Evaluating (4.6b). Finally, students may 'devise and carry out ways of testing the extent to which the product satisfies their design specification'. ACTIVITY: the artefact has to be tested against the original specification as part of the final report to the brewer.

STUDENTS' SECOND ASSIGNMENT AND THE NATIONAL CURRICULUM

Assignment

A Brewer requires a means of sampling a fermenting liquid so that a saccharometer (hydrometer) can be used to measure the amount of sugar remaining in the mix. It is possible that an automatic system of sampling and measuring could be devised, but in all probability students will attempt to produce an improvement on the ladle currently used. Its construction would involve fairly elementary craft techniques in one of several materials which could be used for a successful solution. The students are asked to present a report to you as Managing Director.

Programme of study

- Analyse alternative solutions to produce a better design proposal.
- Use modelling techniques to communicate design proposals.
- Implement systems in which control is maintained without need for human intervention.

Attainment Targets

Identifying needs and opportunities (1.6a). Students should 'explain how they have identified needs and opportunities for design and technological activites and give a justification of the conclusion they have reached'. ACTIVITY: a detailed design specification is required and the content justified for use by the Brewer in presenting a case for adoption to the Managing Director.
Generating a design proposal (2.6a). Students should 'produce a design specification and use it to develop their design proposal'. ACTIVITY: an elementary design specification for the sampler is provided, but this has to be extended and used in the ultimate manufacture of a prototype.
Planning and making (3.5b). Students must 'use knowledge and understanding of the properties of a range of materials in their planning and making'. ACTIVITY: in this process, hygiene is of particular importance and so the material for the system must be workable into an artefact that can be easily cleaned.
Evaluating (4.5b). Students are asked to 'evaluate their product in relation to the design intentions and to the original needs or opportunities'. ACTIVITY: an evaluation must be included in the final report to the Brewer.

Topic networks

Students could be requested to devise topic networks and in so doing to develop an understanding of the relationships that this industry has with its environment. The one shown here relates simply to materials. Others could be devised that concentrate upon such topics as workers, techniques, places. This should help students to meet parts of Attainment Target 1, level 5, which states that they should be able to 'recognise that economic, social, environmental and technological considerations and the preferences of users are important'.

Materials network

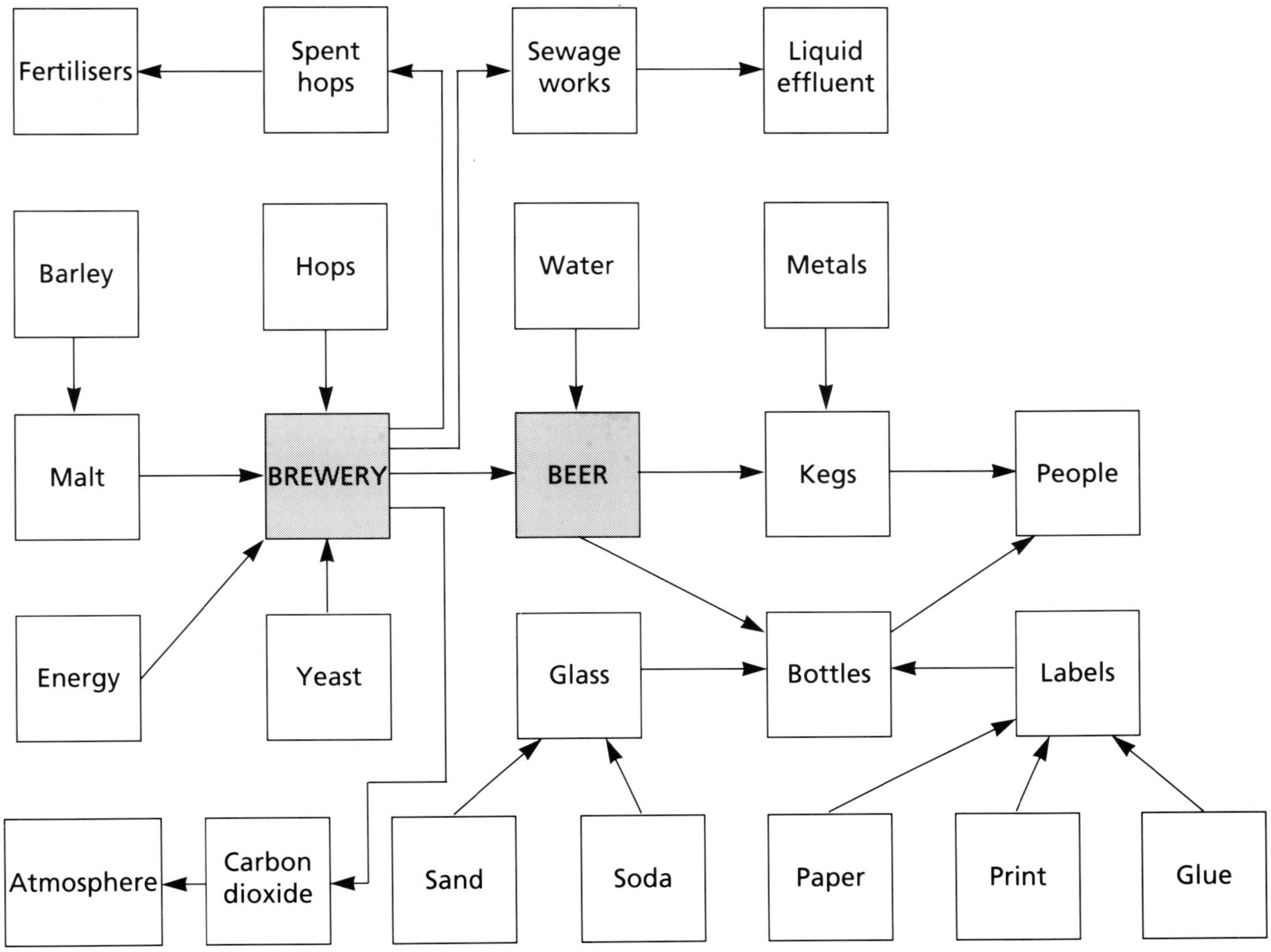

THE INDUSTRIAL PROCESS – A CONTEXT FOR PROBLEM SOLVING

Why drink alcohol?

Drinking too much alcohol can kill you! Nevertheless, in a year enough beer is drunk in the UK to average out at 110 litres (about 180 pints) for each inhabitant.

So why do people drink beer? Some enjoy the feeling of relaxation that a drink or two gives them, though some become depressed when they drink. Some drink to impress others, usually without much success; many just like the taste. Whatever the reason, there is certainly a demand for good beers and has been for hundreds of years. The Romans brewed ale from malt, after germinating and then roasting seeds of the barley plant.

The first tax on ale was collected in 1188 by Henry II and so the brewing industry came under close scrutiny and control. Even then there were Revenue Officers who checked the quality and quantity of a brewer's efforts, but only to be sure the duty was paid in full. There was little concern with the public's health and safety. Beer did not become popular and readily available until much later.

A brewer uses yeast to convert sugar by fermentation into a solution of alcohol. Hops give the additional flavour and bitterness which characterise beer.

There are many different kinds of beer. Their strengths, decided by the amount of alcohol they contain, are all different. In comparison to whisky (about 40% alcohol) and sherry (about 20%), beers are quite weak: even an extra strong beer contains only about 9% alcohol. Weaker beers contain about 4% alcohol.

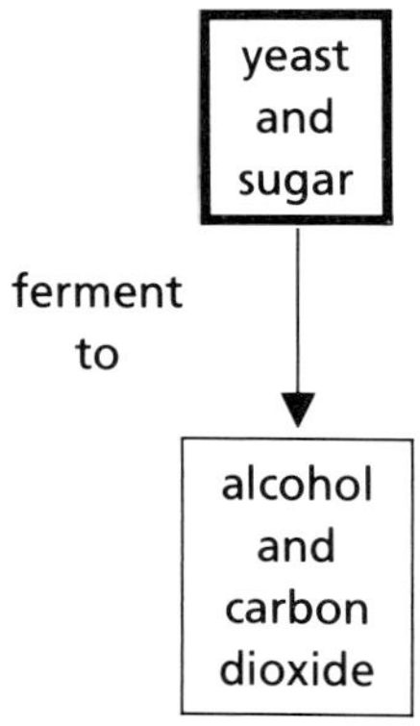

Excise duty and the Revenue Officers

To pay for such services as health, education, fire, ambulance and police, the Government needs money. This revenue is raised through taxes, such as income tax and value added tax (VAT), and duties. There is a duty that must be paid on all beers that a brewer produces. The amount to be paid is calculated by officers of the Government's Board of Customs and Excise.

Revenue Officers often inspect what is being brewed. Some large breweries have their own Revenue Officers based permanently on-site.

A problem faced by the Revenue is that a brewery will not produce the same amount of beer each day, and different brews vary in strength. A regular check is needed. The strength of spirit was at one time measured by pouring it onto gunpowder and setting the mixture alight. The rate at which the gunpowder burned was an indication of strength. Another test was to pour oil onto the spirit and see how well it floated. This method is similar to the one used today to measure the strength of beer.

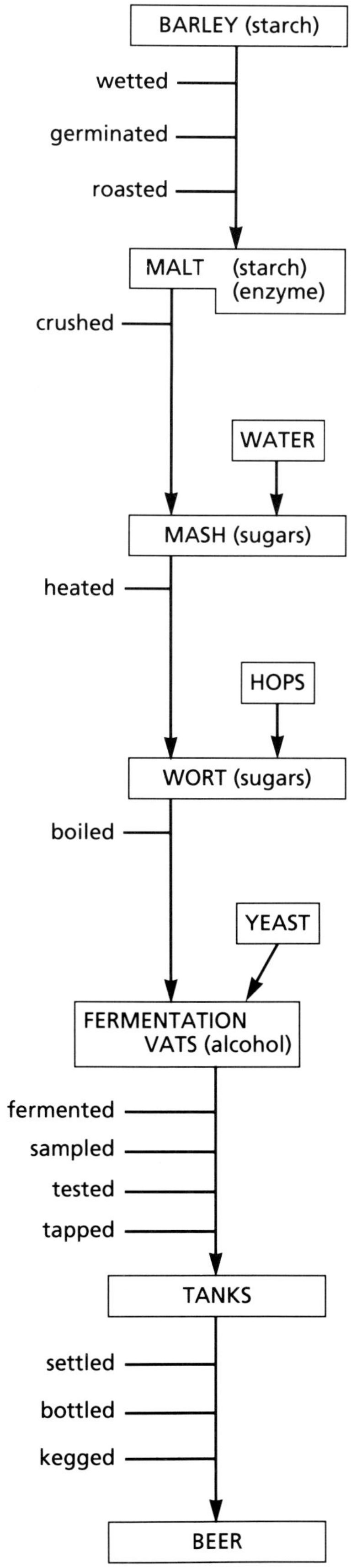

The Brewer's task

Put simply, a Brewer changes flavoured sugar solutions into flavoured alcohol solutions. The sugar comes from the starch present in barley grains. These are soaked in water for several days until they begin to sprout. During the sprouting an enzyme is formed which helps the change from starch to sugar.

The seeds are heated to stop any further growth and to improve the flavour of the beer. They are then crushed and heated with more water. In this way the barley's starch is changed by enzymes into flavoured sugars, mainly maltose.

This sugary mixture is boiled for two hours with the flowers of the female hop plant to give extra flavour and bitterness. The spent hops are then separated and sold as fertiliser.

After cooling, yeast is added. Over three to five days it feeds on the sugars and converts them into alcohol and carbon dioxide. At this point the liquid needs to be sampled to test how much sugar is left and how much alcohol is present.

The alcoholic solution is then tapped off into tanks. On clearing, the beer so formed is kegged or bottled. The yeast can be used again.

PROBLEMS FOR BREWERS AND ASSIGNMENTS FOR YOU

To enable the duty to be paid, sufficient beer must be sold. Increasing the range of beers for sale helps by appealing to different tastes. For good sales it is essential to solve the problem of how to tell potential customers about the quality of new products. However, if the beer is not up to standard, then a company can go bankrupt. Another problem therefore lies in taking samples of the fermenting liquid and ensuring that it is of the correct quality.

Now read a Brewer's description of his job. He will suggest two assignments that you could attempt.

A Brewer's work

My name is Philip Brunt and I'm a Brewer working for Gibbs Mew in Salisbury. Salisbury is a good place for a brewery as there is plenty of water, essential for making beer.

I hope your idea of a Brewer is not an elderly gentleman who relies entirely on years of experience and is surrounded by weird bubbling mixtures that need constant tasting to see whether they are ready! There are lots of bubbling liquids around me when I work, but brewing has much science in it as well as being an art. My training certainly helps. I have a degree in biochemistry from the University of Bath and a master's degree in brewing science from the University of Birmingham. All the same, you would be right about needing a good sense of taste and smell. I need both, to judge the quality of the beer I brew and the suitability of all the malt that arrives at this brewery.

With a wife and two young children, I feel fortunate that I don't have to work shifts. Even so, the production of alcohol goes on night and day as I have millions of workers that keep the fermentation process going without stopping and don't expect to be paid, except for a lot of sugar. I expect you have realised those workers are microscopic yeast plants, which take in sugar and give out alcohol, with some carbon dioxide.

My main task is to be sure that brewing proceeds smoothly. This involves checking that our human workers are content and doing their jobs and that the yeast is warm and well fed. We are also continually looking for new products. One of them could be a new drink called 'Bishop's Brew'. You may be able to give some indication of how best to bring this beer to the notice of potential customers.

Another concern I have is with the regular testing of our brews. Our system of sampling is fairly simple and while it seems to work quite well I think it could be improved. Some help here would also be useful. So – read the next few sheets, put on your thinking caps and be as industrious as my yeast!

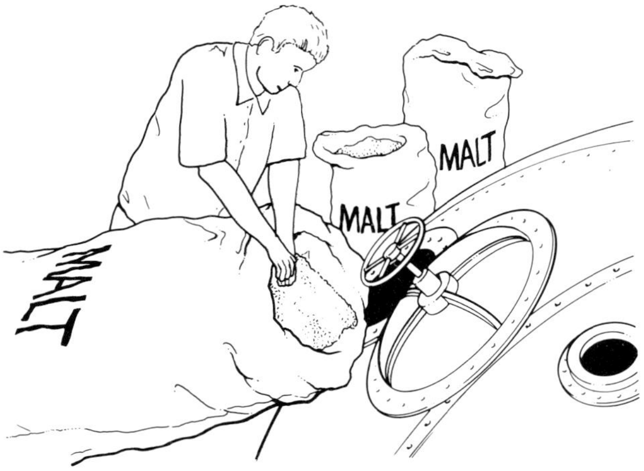

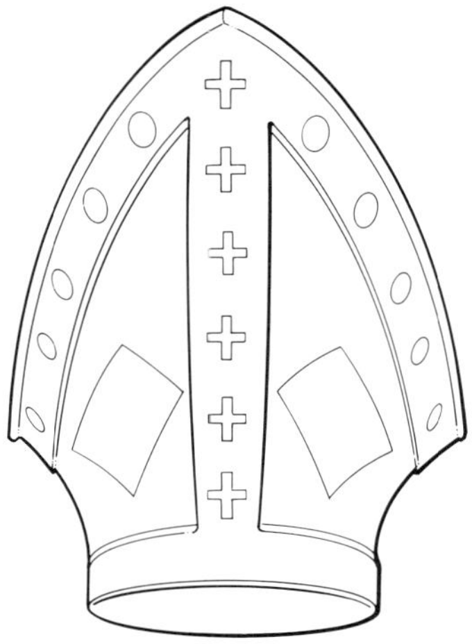

The Brewer's first project – advertising a new beer

I brew a special beer that up to now has been supplied only to some clergymen. I call the beer 'Bishop's Brew'. This year I'm going to brew much more for sale to local pubs. I need some help with advertising.

> *Information – Bishop's Brew*
>
> *The beer is a straw-coloured, delicious liquid. With a strong taste of hops, it's classed as a 'bitter'. It is to be sold as a draught bitter in wooden barrels.*

I'd like some advice. First, what do you think of the name? Can you think of a better one? Don't forget we are based in Salisbury and our name is Gibbs Mew. Should there be a logo? Secondly, I need an advertising campaign. What form should this take? I have had one good idea but I need some more. Give me examples of several ways in which we could advertise.

My idea is to advertise the beer by giving away beer mats as an advertisement. But what form should these mats take?

> *Information – beer mats*
>
> *Breweries give away beer mats to protect polished surfaces from spilt drinks. The mats can be made from all sorts of materials, with different designs, colours and shapes. Some people collect sets of such mats.*

I have had the following thoughts about these mats. Nevertheless, as you are the designer, I expect you will have others.

Should the material absorb liquids?

What size/ thickness/ shape/ number?

Should words be included?

What illustration– serious/ humorous/ coloured/ in sets?

Assignment 1 – an advertising campaign
Please write an illustrated report, including:

- your thoughts about my suggested name for the beer
- a list of the ways in which you believe the beer should be advertised
- a design for a set of beer mats that I could use as part of this campaign.

I hope that in the section on the mats you will include a design specification and a description of how you have tested your product (or its mock-up) against that specification.

The Brewer's second project – sampling Bishop's Brew

In the vats, sugar is changed into alcohol by yeast. Alcohol makes the beer intoxicating. The sugar and flavour comes from malt. Hops give the beer its bitterness. I need to know how much sugar is left in my vats and I need samples of the fermenting mixture for testing. This sampling is causing me problems: I need a better method.

To estimate the sugar, I need to measure the specific gravity (I expect you would call it relative density) of the beer. I collect about a litre of liquid from the vat using a large stainless steel ladle. Its length, about a metre, makes it awkward to use and so the work makes my back ache.

I need a sample from about a metre down in the brew and away from the side of the tank. I must be careful not to collect any of the froth, yeast and hops that float on the top.

When the ladle is full, I rest the handle on the side of the vat to take some of the weight and put in a saccharometer (you probably call it a hydrometer) which is about 30 cm long. The height at which this floats gives me a measure of the sugar left and the alcohol present. I record the measurement.

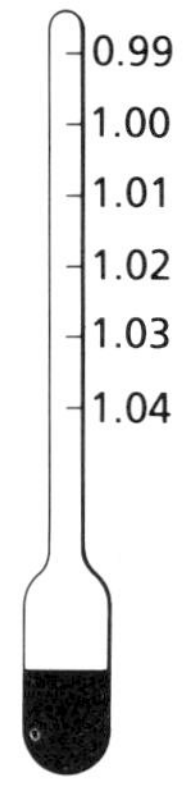

Displaced beer overflows the ladle and runs back into the vat. Any falling on the floor would 'go off' and smell. Finally, I pour the sample back into the vat, sterilise the ladle and store it away.

I have to do this several times a day. Now read my notes and make a suggestion for a new sampler, including instructions for its use.

These are the notes the Brewer made about his sampling system.

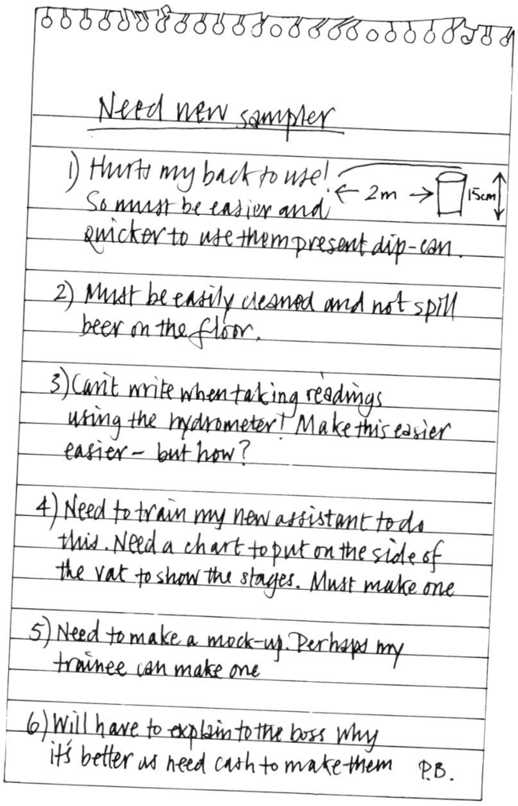

Assignment 2 – a system for sampling

Use the Brewer's notes and what you have read previously to produce a detailed design specification for a dip can. Then design, make and test it. An instruction sheet is also required for trainees using your new system.

Produce a guide for the Brewer to use in planning a presentation to his Managing Director that includes a justification for your final design specification and how your prototype has been evaluated against this. Play the role of the Brewer and present your design to the Managing Director; your teacher will probably agree to play this role.

High Street stores – creating images

Imagine yourself about to walk down the High Street of your nearest large town. The chances are that you will pass a series of windows that display much more than what each trader has for sale; the windows really portray the image that the store has of itself.

If Display Assistant, Lisa Boswell, and her colleagues are doing their jobs well, the Marks & Spencer windows will catch your eye with an image that portrays value, freshness, style, colour and reliable high quality.

Lisa sets two assignments. The first involves a survey of the buying habits of men and women. The second requires the design of a display for a 'penny bazaar'.

BACKGROUND INFORMATION

The Student Information describes the work of a Display Assistant at Marks & Spencer. Giving advice about means of displaying merchandise of various types, it places the assignments described below in context. These assignments can be completed independently of one another. The second benefits from several groups or individuals working separately so that their displays can be evaluated against the factors considered important for a display of the merchandise chosen.

STUDENTS' FIRST ASSIGNMENT AND THE NATIONAL CURRICULUM

Assignment

Ways in which the buying habits of men differ from those of women are suggested and the worth of these is estimated by students through some form of market survey. Recommendations are required from students about how the results might affect the way in which merchandise should be positioned for sale and display.

Students will have to decide whether to work in groups or as individuals and will be required to present their results to a Store Manager, a role perhaps played by yourself.

Programme of study

- Allocate tasks when leading a team.
- Present proposals to an audience, using a range of methods and media.
- Collate, sort, analyse, interpret and present information in a logical and coherent way.
- Review the ways in which market research can be used to evaluate user requirements and market potential.

Attainment Targets

Identifying needs and opportunities (1.8b). Students are required to 'plan in detail the various stages of their investigation'. ACTIVITY: pupils develop their own plan, strategies and questions for data collection.
Generating a design proposal (2.5e). Students must 'specify what they intend to do and what they will need by using simple plans and flow diagrams'. ACTIVITY: the strategy for data collection should be considered in depth before commencement.
Planning and making (3.6e) requires students to 'show judgement in seeking advice and information'. ACTIVITY: the quality of the information obtained is dependent upon the quality of the questions asked, the methods used to elicit the information, and the effective use of time.
Evaluating (4.6d). Students should 'evaluate the procedures, techniques and processes used and indicate possible improvements'. ACTIVITY: pupils review their market research strategy and evaluate the data obtained in their final presentation to the Manager.

STUDENTS' SECOND ASSIGNMENT AND THE NATIONAL CURRICULUM

Assignment

Students are asked to design a display for a 'penny bazaar'. The choice of product and display material is left to each student. Only five articles in total may be used in the display, and a chair must be used as a prop. It is intended that more than one student will attempt this assignment and so some comparison can be made of criteria for success that have been drawn up by students.

The success of displays can be judged by 'customers', a role perhaps played by the remainder of the class.

Programme of study

- Use materials economically and efficiently.
- Develop craft skills.
- Design the appearance of an artefact, system or environment so that it appeals to users.

Attainment Targets

Identifying needs and opportunities (1.7c). Students should 'consider both the user and the producer when defining the need for a technological activity'. ACTIVITY: the design of a stand for the sale of merchandise brings the user and the designer together. The designer cannot ignore the perceptions of the user with regard to the stand and its products.
Generating a design proposal (2.5d) requires students to 'establish and check the availability of the resources required, adapting their design as appropriate'. ACTIVITY: the design of the stand is likely to be determined by that which is to be displayed and the supporting display material.
Planning and making (3.6b). Students should 'combine knowledge of the properties of a range of materials and of processes to identify those most suitable for their design'. ACTIVITY: the stand is likely to involve different materials which are chosen for different reasons.
Evaluating (4.6c). Students must 'evaluate the ways in which materials have been used'. ACTIVITY: the effectiveness of the design will be determined largely by the imaginativeness and appropriateness of the materials used.

CHAIN STORES ON VIEW

It is over 100 years since a pedlar called Michael Marks set up a stall in Leeds market, but it was another ten years before he was joined by Tom Spencer to form the partnership that resulted in the chain of shops that bears their names. The chain is now considerable, with over 290 stores in the UK and Eire and with an annual turnover there in 1990 of nearly five billion pounds. You might like to calculate the average amount per store. In 1990 Marks & Spencer employed more than 20,000 people full-time and over 40,000 part-time. The future is one of expansion both in the UK and overseas, where additional developments are planned for Europe, Canada, the USA and Asia.

This chain store has developed an image of good quality, value and reliability. The maintenance of this reputation is vitally important to its continued success. The first thing that potential customers see, when walking down the High Street, is a display of merchandise in a series of shop windows. Once a customer has stepped onto the shop floor, the image must be further enhanced. This is why the work of the Display Assistants is so important. They continue to develop that image through the store's general decor.

Lisa Boswell is a Display Assistant at a large High Street branch of Marks & Spencer. She agreed to tell you about her job and how she spends her working time. Just before she began, there was a noise from her 'bleeper' and a voice came through asking her to phone an internal store number. Lisa heard that

eager customers had taken several unwrapped items from a display, spoiling the design that she had created only that morning. Lisa was quite understandably dismayed, especially as identical, wrapped items were close by and within easy reach. That, she said, is one of the hazards of making displays too attractive.

Designing and presenting displays

I'm Lisa Boswell. This is my first year with Marks & Spencer. Last year I obtained a BTEC National Diploma in Design and Display at our local college and, soon after, I was fortunate enough to be employed in this large store as a Display Assistant. It really is an enjoyable job, even though I am very busy. I'm the only full-time Display Assistant at the store. I work a five-day week, with alternate Saturdays off. My hours are regularly from 8:50 to 17:40, with an hour for lunch and twenty minutes each for two additional breaks, one in the morning and one in the afternoon. I have over five weeks' holiday each year.

This store has three floors for the sale of our merchandise, and another for staff offices and amenities. These include a hairdressing salon and a medical room where a doctor and dentist visit during the week. A chiropodist comes in once a week, as staff are on their feet for most of the day.

Our Head Office specifies how much floor space has to be given as an operating area for the different categories of merchandise, for customer service desks, and for walkways. However, because the stores are all shapes and sizes, they all have different layouts.

In our store we have five main sections: menswear, ladies' wear, children's wear, food, and fabrics and furnishings. Menswear and the food hall are on the ground floor, with ladies' wear on the first floor and fabrics and furnishings and children's wear on the second. Apart from the food hall, which is best placed on the ground floor for easy access, this positioning just depends upon how best the sections fit the space available in the building.

For the Christmas period, I will be involved in setting up what we call the 'penny bazaar', which displays merchandise that is suitable for presents. This section has to be positioned for easy access.

My job is to design and present displays. I hope you don't think I spend all my time dressing windows. In this large store there are only three exterior windows and they are by no means large. Each is redesigned every month and that takes me about four days in total. The rest of my time is spent on displays within the store – and that is the busy part. As you will see when you next pop in, there are displays all over the place.

Of course, I don't have complete freedom as regards what is displayed or how it is presented. With such a large chain of stores there is a need for a 'corporate image' to be portrayed by all stores. To achieve this I receive additional training and instructions from M & S Head Office.

Now I'm going to pass on to you some of my ideas. You may want to make use of these when you start on the two assignments I've set for you at the end of this section.

In all my designs I need a balance between business and art, and to help me I have certain props. I'll give you details of some of these, though we have lots of others.

A simple chair

It is exciting to find that a lot of merchandise can be displayed on a chair, and not only clothes! Remember that people's eyes will flow down the design, so be sure that the line is clear and unbroken. Accessories such as belts and shoes should complement your design.

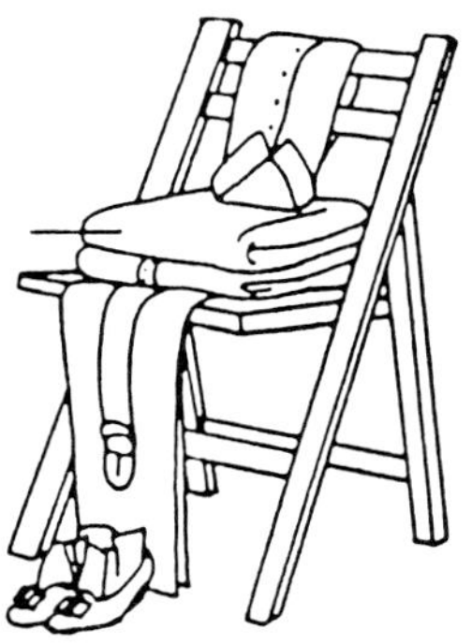

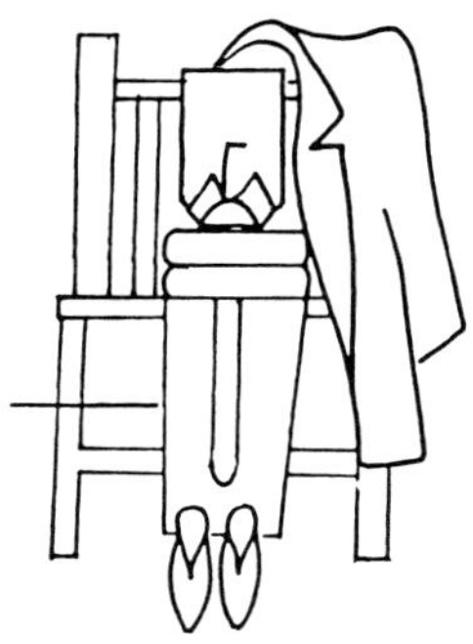

Valet unit

Mannequins, or dummies, are not the only way to display a full set of garments in a natural position. The valet unit, with its variable height, has much potential especially for small windows.

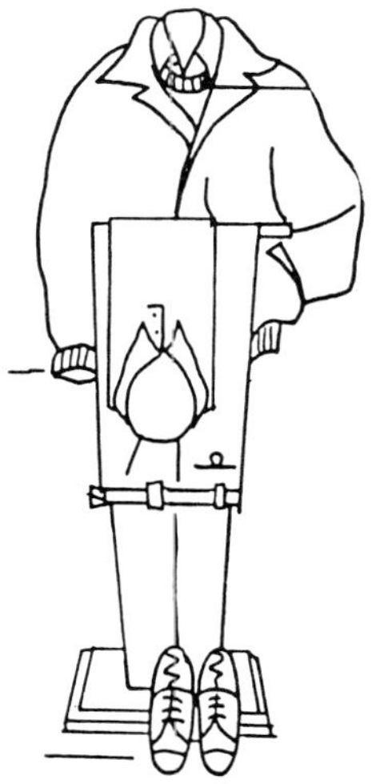

Keep casual and formal merchandise separate

Keep to your colour scheme

Keep to one theme -gardening/beachwear/party/dance/school

For a large window I must consider what merchandise to select and how to create balance. Of course, I need to liaise with Departmental Managers in deciding what is to be displayed, incorporating the colour scheme specified by Head Office.

What I am really doing every time is putting together a series of coloured shapes and trying to group these effectively. The complete design should form a triangle. If a design is to be symmetrical about the centre line, the overall shape should form a symmetrical triangle. If the design is not to be symmetrical to the centre line, the shape of the triangle should be pushed to one side.

I first decided what is to be the focal point of the design.

I remember the invisible centre line that will run down the middle of the design

I know that I can have a balanced design with or without symmetry about this centre line

However, if the design is without symmetry then the focal point should be to the left, or right, of the centre line

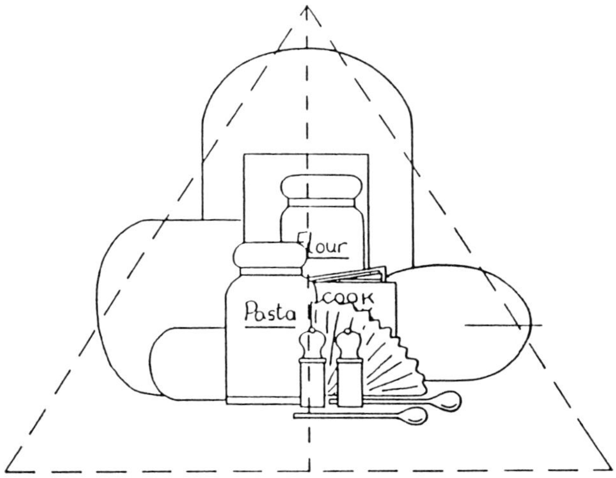

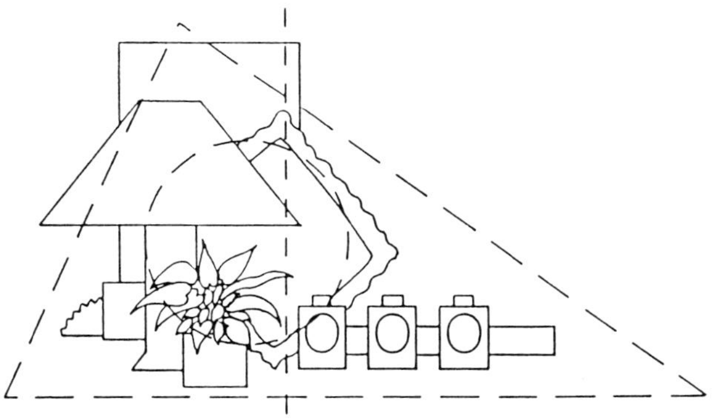

Do you think one of these similar designs has a better balance than the other, or are they equally well balanced? Why do you think so?

TASKS FOR DISPLAY ASSISTANTS

The Display Assistant's first project – do men have different buying habits from women?

The objective of the store is to maximise sales. To do this we try to make the store like a shopping arcade with its own walkways and groupings of products. Displays are placed in prominent positions to catch the eye and generate interest.

In considering where to position products and displays, we carefully observe the buying habits of men and women. I wonder whether there is any truth in what I have read about men and women buying clothing!

70% of menswear is bought by women for men

For women, shopping is a pastime to be enjoyed and time does not matter

For men. shopping is a chore to be completed as quickly as possible

On entering a store, women have no fixed idea of what they want

On entering a store, men know the colour they want and only look for the style

Assignment 1 – do men have different buying habits from women?
The Store Manager needs evidence about the shopping habits of our customers. Your task is to carry out a survey that will collect information which will help you make judgements and recommendations. You should think about these questions:

- Should you work as part of a team or as an individual?
- How will you organise yourself or your research team to collect this information?
- How will you ask the questions?
- How will the answers be analysed?
- What information do you want?
- What questions will you ask?
- How will the answers be recorded?
- How will the results be reported?

The Manager is extremely busy and you should present your information so that it can be understood easily and quickly. (Your teacher may play the role of this Manager.) The Manager will certainly expect you to make recommendations about how your results affect either how menswear and ladies' wear are positioned in the store or the way in which Display Assistants arrange and position merchandise for display. Nevertheless, you should warn your Manager of any weaknesses in your survey that might invalidate the results.

The Display Assistant's second project – a display for a penny bazaar

Our store is constantly changing. In January it is sales time. In February we sell clothes for spring. At about the middle of June we display clothes for high summer. At the end of July we have sales again. In September there are autumn clothes to show off, and towards the end of November we change our displays for Christmas, which includes a 'penny bazaar'.

At the penny bazaar we display items which customers might buy for Christmas presents. Your assignment is to create a display for your own penny bazaar. However, there is a limitation: you may only use a single chair as a prop on which you may display a maximum of five articles. Of course, the size of the chair will depend upon you. For example, if you decide that your penny bazaar will sell only rings, then your five rings could be displayed on a tiny, model chair.

However, you should consider these questions:

- What items should you sell?
- How can you collect items for display?
- What items, and what theme or colour, should be the basis of your display?
- What information needs to be added to the display?
- How should this information be presented?
- How should the items be grouped?

Assignment 2 – a display for a penny bazaar
Produce a display for a penny bazaar. It could be that other members of your teaching group will produce similar displays. Be sure that you have made a list of the criteria you feel your final display should be judged by. You may want to compare these criteria with those of other students. The rest of your class could play the role of customers and make a judgement about the success of your display.

Postal services – parcels by the million

Imagine the problems, or pleasure, of running a business that in a year collects over 200 million parcels in centres spread through the UK and sees that they are delivered to any address in the world.

All incoming parcels need to be sorted into broad geographical destinations and this is completed at one of Parcelforce's fifteen Parcel Distribution Centres. Jack Williams works at the Bristol Centre, where he is the Control Officer. His major task is to ensure the safe and quick passage of parcels through the Centre on to the next stage in their journey.

How can the thousands of parcels that arrive at the Centre each day from local post offices be distributed worldwide when, on the same day, thousands of parcels arrive from the other major Distribution Centres for local delivery?

Jack Williams explains how this problem is solved and sets two others. The first is to devise a publicity scheme that will encourage public awareness of the need for better packaging. The second is to devise a system that will indicate how full the containers are into which the sorted packages have fallen.

BACKGROUND INFORMATION

The information about Parcelforce is presented by a Control Officer working at a Parcel Distribution Centre. He describes how Parcelforce deals with millions of parcels each week and presents the problem of burst packages. Visits by teachers may be arranged directly with managers in your local Distribution Centre.

The Control Officer sets two assignments. Details of these and how they satisfy aspects of the National Curriculum are set out below.

STUDENTS' FIRST ASSIGNMENT AND THE NATIONAL CURRICULUM

Assignment

Badly wrapped parcels which come apart are a problem at the Parcelforce sorting offices, and students are asked to submit, by means of an illustrated verbal presentation, a poster for display and a leaflet for distribution to bring this problem to the notice of the general public and provide suggestions for a remedy. Copies of a leaflet produced for this purpose entitled 'Wrap up well' are available from Royal Mail.

You may be asked to play the Managing Director for the presentation.

Programme of study

Develop a product (in this case a publicity campaign) and . . . market, promote and sell it.

Attainment Targets

Identifying needs and opportunities (1.7d). Students should 'identify and draw upon sources of expert advice relevant to the identification of needs'. ACTIVITY: the requirement is to produce a leaflet which specifies how parcels can be wrapped efficiently. The instruction is to seek expert advice in doing so.
Generating a design proposal (2.6b). Students must 'produce a design proposal recording their decisions and the ways of reaching their chosen outcome'. ACTIVITY: the scheme has to be 'sold' to the Managing Director of Royal Mail Parcelforce and students must explain precisely how the intended outcome, i.e. communicating with the public in general, will be successfully achieved.
Planning and making (3.6e) requires students to 'show judgement in seeking advice and information'. ACTIVITY: the advice sought in producing the text for the scheme's leaflet should be noted.
Evaluating (4.7a). Students should 'present an evaluation of their activities against the original need, drawing on information gathered about the product and the reaction of users'. ACTIVITY: the structure and form for an evaluation of this nature is required in the presentation to the Managing Director.

STUDENTS' SECOND ASSIGNMENT AND THE NATIONAL CURRICULUM

Assignment

The parcel sorting office has an intricate system for feeding parcels along conveyor belts to sorters. After sorting, the parcels slide into a series of large, 3 metre square bins. The Control Officer needs, while at the control panel, to know when these bins are full. A remote system of detection is required with the aim of developing students' methods of seeking ideas in addition to their own. You or one of the students could then play the role of this Control Officer in accepting students' reports of their work.

Programme of study

Design and implement an artefact containing a system which involves control.

Attainment Targets

Identifying needs and opportunities (1.9b). Students should 'review their own knowledge and draw up a strategy to exploit expert sources'. ACTIVITY: various means of solving the problem are sought and some means of extending these using other sources are required.
Generating a design proposal (2.5a) requires students to 'extend their first ideas by combining various aspects of them to formulate a design proposal and explain why some ideas were not used'. ACTIVITY: the first stage in the project is for students to find ways, using only their own expertise, of solving the problem of detecting when the bins holding sorted parcels are full and to decide which solution is the most appropriate.
Planning and making (3.6e). Students must 'show judgement in seeking advice and information'. ACTIVITY: the second stage is to find, from other sources, additional methods of solution and to decide whether these can add to their original design.
Evaluating (4.6a). Students are to 'evaluate the procedures, techniques and processes used'. ACTIVITY: finally, the students have to explain, using a mock-up, how their chosen system will achieve the intended outcome.

NATIONAL PARCEL DELIVERY

My name is Jack Williams and I am the Control Officer at Parcelforce's Bristol Distribution Centre. Nationally, Parcelforce handles 200 million parcels each year. These deliveries employ approximately 12,000 people, including clerical staff, drivers, parcel sorters, parcel handlers and motor technicians. Our parcels are transported by a fleet of approximately 6000 vehicles that travel in total nearly 80 million kilometres (50 million miles) each year.

We are involved in a giant operation. Like all other businesses, we must make a profit, and cannot do so unless we provide a service that people are prepared to pay for.

Choosing the best service for one of your own parcels

To get an idea of the range of services offered, pretend that you have a parcel that needs delivering. Make a guess at its weight in kilograms and decide where it is to go and when it must arrive. Then decide from the following which of our services would be the best for your parcel.

- Do you need our standard, cheapest service? This delivers parcels weighing up to 25 kg to addresses in the UK and normally takes about three days.
- If your parcel weighs less than 30 kg, one of our two services that ensure delivery in 24 hours or 48 hours, respectively, might be suitable for your parcel.
- Or do you need Datapost, with its overnight delivery and guaranteed delivery times on the following day?
- If you want proof that your parcel has been delivered, you should choose Trakback. This uses the latest computer technology to provide proof of delivery.
- Finally, if your parcel is to go overseas, we have a sea and air service, a special form of Datapost.

A MAJOR ORGANISATIONAL PROBLEM

On a weekday at any Parcel Distribution Centre we have two major tasks that must not become mixed up. One is to deal with parcels that have been collected locally and need sorting before distribution to one of the fifteen national and two international Distribution Centres in the UK. A second task is to sort parcels that are delivered from other Distribution Centres and have to be distributed to addresses within the Centre's area.

At Bristol, we prevent confusion by separating the tasks, not by putting them in different buildings, but by finishing the first sorting task in the time from 06:00 to 15:00 and then changing completely to our second sorting task and completing this from 15:00 to 24:00. Parcels are of course arriving from the local area and from other parts of the country all day. Only the sorting is split into the two shifts.

Parcel delivery

Our work takes place at the point where our road vehicles arrive with their parcels. Don't imagine that they are filled with parcels like pickled onions in a bottle, because they are not. Some parcels are held in sacks, though most are in large hard-sided containers with wheels. These containers have a place for the prongs of a fork-lift truck to fit, so that they can be picked up with ease.

The vehicles vary from small vans to articulated lorries weighing 32 tonnes that can carry up to twenty containers. However, all are coloured red and have our logo on their sides for all to see. These vehicles are the hub of our organisation and they depend upon an experienced and efficient force of drivers. It is a legal requirement that vehicles weighing over 3.5 tonnes when loaded are fitted with a tachometer. This records several measurements, including distance travelled and speed.

We are near the centre of Bristol, adjoining Temple Meads railway station. Once this was very convenient but not now as we do not send parcels by rail. There are thirteen unloading bays, all in line. Vehicles reverse into these bays to unload. The yard is very busy all day with trailers being dropped off and picked up. We also have vans and lorries bringing in and taking out sorted parcels.

With articulated lorries it is possible for a driver from another Centre to meet one of the Bristol drivers somewhere along the journey at a convenient Trailer Exchange Point and swap trailers. This enables drivers and their vehicles, though not trailers, to return every day to their home base – a convenient and efficient system.

Unloading and sorting

As we remain open for five 06:00–15:00 sessions and five 15:00–24:00 sessions during each week, we work shifts. This week I am on 'earlies' but next week I shall be working 'lates'. My task is to make sure our system of sorting parcels works efficiently.

The early shift (06:00–15:00)
When the vehicles bringing parcels in are at the bays, unloading can begin. All sacks of parcels are unloaded by hand but containers, as they are very heavy, are unloaded using fork-lift trucks. The trailers are backed up to a platform, and a conveyor belt which passes under the platform is accessed by pressing a button that opens a trap door. Sacks are emptied directly onto the conveyor belt. The larger containers are rolled to a tipper which lifts and empties them onto another conveyor.

If the parcels travelling on all these conveyor belts were not controlled in some way, chaos would result. I supply the control from a small room that looks directly onto part of our Centre and views all of the remainder with the help of video cameras. From here I can see where there is room for additional parcels, and where the pressure is so great that no further parcels should be sent there.

The parcels passing along the conveyor belts are still unsorted, but at this stage sorting can begin. The variety of parcel shapes is great: our standard service will accept parcels up to 1.5 metres long and addresses can appear just about anywhere on them.

The belt and all its parcels must pass in front of a sorting operator, who must inspect the parcel to find an address and so decide which of the Distribution Centres handles this address. Parcels bound for overseas are sent to one of two international Centres. The belt from the operator onwards is inclined sideways, with parcels being prevented from sliding off by a guide rail. On reading an address for a parcel, the operator presses the appropriate destination button. A computer relays the speed of the belt, the moment the parcel crosses the beam of a photo electric cell, and the distance along the belt to the package's destination chute so that the gate on the side of the belt falls open just as the parcel arrives. It slides along a chute into a storage bin for further sorting and checking.

When all 'incoming' parcels have been sorted into containers, they are loaded onto one of the vehicles and taken to the appropriate Distribution Centre. We aim to clear all incoming parcels by 15:00 hours.

The late shift (15:00–24:00)
The procedure for handling traffic on the late shift is identical to that of the early shift except that the sorting is done for those parcels arriving from other Distribution Centres and bound for local addresses.

The parcels sorted during the early shift are now being loaded onto vehicles for despatch to local Centres. On their return journey they will bring parcels posted in those local Centres' areas.

The Control Officer's first project – encouraging efficient packaging

Many parcels come apart during transit because of the way in which they have been wrapped. This is such a problem that some members of my staff spend all their time repackaging and readdressing those that have burst. Sometimes the part of the parcel the address is attached to comes loose and is lost. In these cases, my staff have to search the package to try, often without success, to find the address of the sender. Remember that all sorts of goods are transported through the post, including liquids, musical instruments, live maggots, live pigeons, butter, fish and flowers.

We are considering the introduction of a nationwide poster and leaflet campaign to improve the wrapping of parcels that come through our postal system. Imagine that the Post Office is considering the introduction of this campaign and you would like to gain the contract for doing so. You will have to devise a scheme and 'sell' it, in competition with other prospective Campaign Managers, to Royal Mail Parcelforce.

Presentation to the Managing Director
You will be required to make a presentation of your scheme to the Managing Director, a role played by your teacher or perhaps even one of your classmates. If there are others in the class who also want the contract, there may be some competition.

A flow chart of your scheme, including possible timings, is essential. This should show exactly what you would do, should you be awarded the contract for completing the project. You should also display appropriate mock-ups of poster and leaflet, and you will need to justify the content of your leaflet, referring by name to the sources of the advice you have taken. All designs will speak for themselves. Assume that there are no restrictions, other than those of practicality, imposed by the Post Office on the sizes of either the poster or the leaflet. The Post Office is seeking your guidance but will view it critically.

When designing poster and leaflet you will have to consider the impact, clarity and content of the message you are trying to communicate. The positioning of the posters and the distribution of the leaflets are of considerable importance.

Assignment 1 – encouraging efficient packaging
Before preparing your presentation you should do the following, at least:

- Design, including the production of a mock-up, a poster that will make the public aware of the problem of package damage and direct them to a leaflet which gives hints on how to wrap parcels efficiently.
- Design and specify the content of a leaflet that will be sent to members of the public on request. It must contain hints on efficient ways of addressing and packaging parcels, large and small, so you will have to put in some research to decide what these hints might be. You should seek expert advice!
- Specify typical sites, giving examples from the area in which you live, for the placement of the posters so that they receive maximum publicity. This recommendation should include examples of the cost of putting up posters. Remember that fly posting is illegal!

The Post Office will require some measure of how successful the programme has been in achieving the intended outcome. How you intend to make this assessment must be included in your presentation. Questions which answers should be sought for include:

- Have posters been noticed?
- Have people sent for leaflets?
- Has packaging improved?

The Control Officer's second project – knowing when storage bins are full

Our sorting system is operated manually; sorters sit by a conveyor system along which comes a steady stream of parcels. I can adjust the volume of that stream from my control box. Sorters find and read the address on each parcel and punch this destination onto a computer keyboard. The system's computer then allows a period of time to elapse, equal to that taken by the parcel to travel along the conveyor belt to the correct destination gate. Just as the parcel arrives there, the gate opens and the parcel slides down into its storage bin, for additional sorting and checking. Later the parcels are loaded onto a lorry for transportation to their appropriate destination.

One of the original problems, though we think we have solved it, lies in the need to ascertain from the control box when storage bins are full and can no longer take more parcels. With this information, I can then arrange for the bins to be emptied or replaced.

The bins in this sorting office are quite large: 3 metres × 3 metres × 3 metres in size. The problem we have with this is that we cannot see inside them from the control box to tell whether they are empty, partly full, or full. We want a system which will provide this information, taking into account the fact that parcels come in all shapes and sizes.

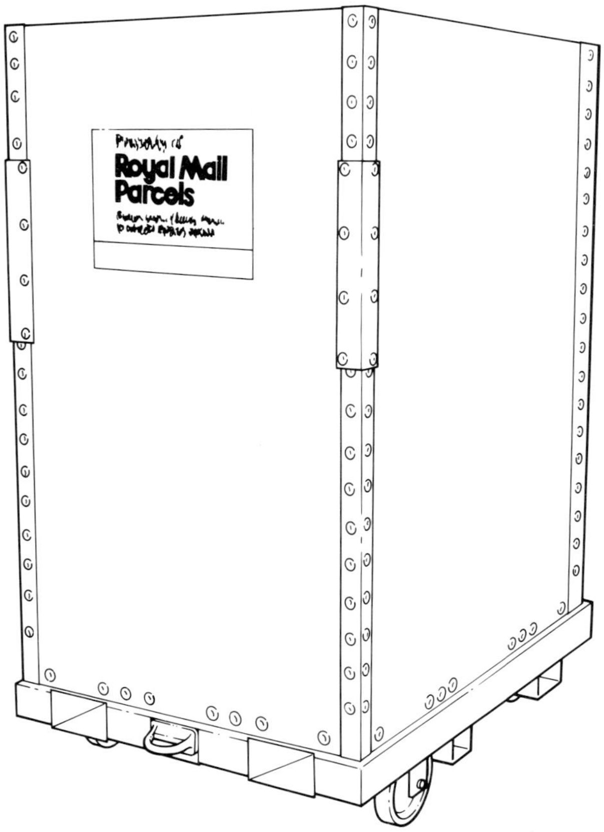

Assignment 2 – detecting when storage bins are full

We need a system for detecting the level of parcels in each of these bins without altering the shape or material of the bin. Think of as many different ways of solving this problem as possible, making a sketch and some notes for each.

Then devise a plan for finding alternative solutions. Seek advice! Decide whether these additional ideas offer an improvement on your solution.

Finally produce a model or mock-up of the best idea and be prepared to justify your choice to a Parcel Distribution Centre Manager, a role which may be played by your teacher.

Research – helping the handicapped

We probably all know at least one person who suffers from arthritis. Osteoarthritis, one of its many forms, affects about half the UK's population by the age of 60. By the age of 70, about 90% of the population is affected by the condition in some way or other.

Arthritis means 'joint inflammation'. In severe cases, a person with osteoarthritis in the knee or hip can become quite unable to move because of pain. However, for some years it has been possible to remove the diseased joint and replace it with an artificial one made of plastic and metal, fastening it in place with bone cement. The success of this operation is high and many people gain a new lease of life when they find that they are again able to walk without discomfort.

Jim Weston is a Technician at Bath University where he is a member of a small team of researchers investigating the problems of joint replacement. He will explain his work and set two assignments that involve other ways of helping those with disabilities. The first requires the design of a telephone dial which a blind person can use with ease, and the second seeks an aid for opening doors for someone with arthritic hands.

BACKGROUND INFORMATION

Both these assignments are intended to show how large research institutions like universities include in their major research programmes a search for solutions to what might appear to be quite small, though important, problems. Research staff, research students and technicians often work together in small teams, with each member offering a different but valued skill.

The handicaps described in these assignments are evidently distressing to the people who suffer from them. By engaging students in these tasks you can help them to appreciate the difficulties of others and realise how Design and Technology has much potential in helping people with disabilities.

STUDENTS' FIRST ASSIGNMENT AND THE NATIONAL CURRICULUM

Assignment

This is based on the case study of a woman who is blind and needs some sort of aid to help her use the press-buttons of her telephone to speak to her son and, in an emergency, to contact her local ambulance service.

Students should work in small groups for this assignment. Other class members could play the role of the Charity Committee that is to decide the value of the artefact(s) produced. They should make an evaluation with the design constraints of simplicity, efficiency and low cost in mind.

Programme of study

- Allocate tasks when leading a team.
- Join materials in permanent forms.
- Have a working knowledge of the properties of a range of materials.
- Use materials economically and efficiently.
- Give attention to detail and work accurately.
- Use modelling techniques to communicate design proposals.
- Recognise the needs of individuals and groups from different backgrounds, when designing for their needs.

Attainment Targets

Identifying needs and opportunities (1.7c). Students should 'consider both the user and the producer when defining the need for a technological activity'. ACTIVITY: pupils are asked to consider the difficulties faced by people with visual handicaps.

Generating a design proposal (2.5c) requires that students 'seek out and organise information to help them develop their ideas and refine their design proposal'. ACTIVITY: only limited information is given; pupils are expected to collect and analyse information relevant to the problem.

Planning and making (3.6b). Students should 'combine knowledge of the properties of a range of materials and of processes to identify those most suitable for their design'. ACTIVITY: a design solution must be specific to the type of telephone referred to in the assignment and is likely to engage pupils in consideration of a variety of materials.

Evaluating (4.6d). Students must 'evaluate the procedures, techniques and processes used and indicate possible improvements'. ACTIVITY: the effectiveness of the solution is to be evaluated against 'effectiveness', 'simplicity' and 'cost'.

STUDENTS' SECOND ASSIGNMENT AND THE NATIONAL CURRICULUM

Assignment

An elderly man suffers from severe arthritis. The assignment requires the production of an aid that will enable him to turn a doorknob easily, so that he can let in visitors. This should involve pupils in an assessment of what, if anything, is currently on the market. The assignment can be used to engage pupils in the whole design process and could culminate in a presentation of their ideas as a report to the Research Group Manager, a role you could play.

Programme of study

- Join materials in a permanent form.
- Give attention to detail and work accurately.
- Identify the advantages of the tools and processes they use for particular purposes.
- Collate, sort, analyse, interpret and present information in a logical and coherent way.
- Recognise the needs of individuals and groups from different backgrounds, when designing for their needs.
- Recognise how economics affects design and technological activities and how important it is to work to a budget.

Attainment Targets

Identifying needs and opportunities (1.7a) requires students to 'analyse information of several kinds and draw conclusions about the needs and opportunities for a design and technological activity, recognising and resolving conflicting considerations about what is worth doing'. ACTIVITY: pupils are asked to consider the range of doorknobs on the market and resolve the issue of whether to design a universal or specific solution to the problem.
Generating a design proposal (2.7c). Students should 'apply relevant criteria including user requirements, costs, time, skill demands, scale of production, and aesthetic considerations, to take decisions about details of the design proposal'. ACTIVITY: the assignment should engage pupils in a rigorous analysis of their design proposal before presentation to the organisation that is providing their funds.
Planning and making (3.6a). Students must 'plan and organise making in order to achieve the desired outcome'. ACTIVITY: pupils are asked to incorporate plans for manufacturing the artefact(s) they have designed.
Evaluating (4.6d). Students are to 'evaluate the procedures, techniques and processes used and indicate possible improvements'. ACTIVITY: pupils present their ideas to others for evaluation of the recommended plans and strategies for construction.

FRONTIERS OF LEARNING

I'm Jim Weston and I work as a Technician at the University of Bath. I'll tell you a little about the university but much more about the research I am involved in.

Universities are places where people study for high academic qualifications such as degrees, and where important investigations, or research projects, are carried out into a wide range of problems.

This university is divided into several Departments, called Schools, which mostly work independently of one another. The one in which I spend my working time is the School of Mechanical Engineering. Including myself, there are fifteen technicians working permanently, full-time in the School. Seven others are employed on a temporary basis on research projects that have funding for short periods of a few years.

Supporting the teaching of undergraduates

About half my time is spent with undergraduate students. My School accepts about 120 students into the first year of their studies for a degree. Young people come here from all over the world but most are from the UK. The qualifications needed for entry are high: about 10% have an appropriate BTEC Certificate or Diploma and the remainder have good grades at A level.

I have to service the workshops in which these students have practical classes. I also provide suitable worksheets, set up demonstrations that lecturers require, and carry out general repairs.

However, the most demanding part of my job is my involvement with what we call the 'two-term design projects'. This is an activity that each undergraduate has to complete: there are over a hundred students in each year and a first degree course takes at least three years to complete. There is a lot of work to do, as many students come to me with their drawings. That's where my experience, skill and general creativity is needed if I am going to be of any help at all in producing a prototype or working model. Project time is a very busy period for me, especially when students mostly 'would like it yesterday'.

Supporting research work

Besides teaching there is research. This is sometimes funded by the Government and sometimes by industry or commerce. It could be that a company has a technical problem and realises that a considerable amount of time will be needed to find a solution and that there is not the expertise within the company to complete the task. Then they could commission members of the university's academic staff to attempt the investigation. These academics will then either complete the work themselves or employ postgraduate research students to work on the problem under their guidance. On successful completion, the company receives a solution to their problem; the student could gain a master's degree or a doctorate.

The School of Mechanical Engineering's areas of research include aero-engineering, automotor engineering, hydraulics and structures. There are smaller projects and I am part of one that is researching into aspects of bioengineering.

Research work is often completed in teams, sometimes large and sometimes small. A team I am working with at the moment contains only three people: a member of the academic staff, a postgraduate student (actually a surgeon) studying for a higher degree, and myself.

A research topic

One of our interests is the bone cement used to fix artificial joints in position. To understand the importance of this work you need some knowledge of human joints and their replacement.

Without a stiff, bony skeleton each of us would collapse in a heap like a jellyfish. But an inflexible and jointless skeleton would be just as impossible, because we would be unable to move at all.

Wherever two bones meet, there is a joint. Not all joints move and our interest is in those that do. Some can move in all directions, such as those in the shoulder and hip. These joints are like those that hold many car aerials in position. Other joints only move in one direction, for example those in the knee and elbow. They are like the hinge of a door. On the end of the bones in all these movable joints is a protective gristle, called cartilage, which is strong and elastic.

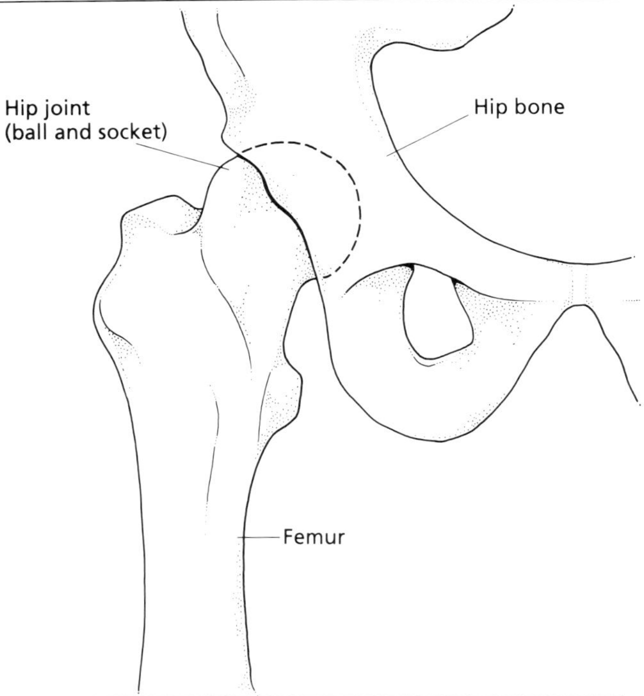

Muscles for movement

Of course, without muscles we would still be unable to move. Muscle may or may not be under control. For example, most people have no control over the rate at which the muscles in their hearts contract to pump blood around their bodies, but most people have good control over the muscles in their legs that enable them to walk.

Put very simply, when two bones form a structure connected by a hinged joint (see illustration), the two bones are connected across the joint by muscles. When the muscles contract, the structure bends at the joint – in this case at the knee. The movement of this joint is lubricated by a special fluid.

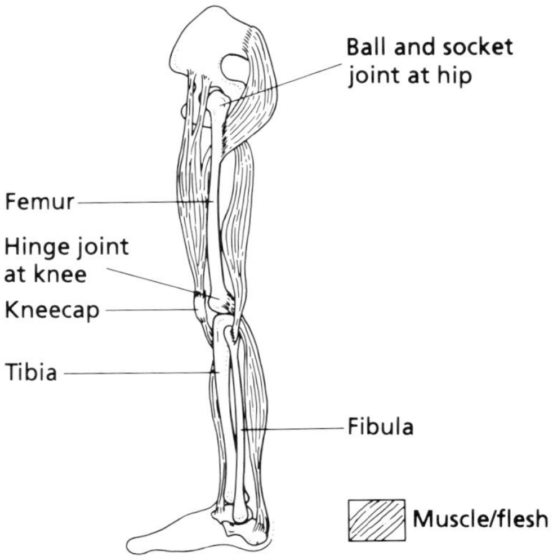

Damaged joints

Joints can be damaged in several ways: through road accidents or sport injuries and by general wear and tear over a period of many years. I will use as an example the ball-and-socket joint in the hip. The ball-shaped bone at the end of the femur turns in a socket in the hip bone. The illustration shows that general deterioration to the bone through age must make any movement of the ball in its socket considerably painful. The pain can cripple people with this condition, making it quite impossible for them to walk. Now there is an operation that can be performed which will often successfully replace the joint completely.

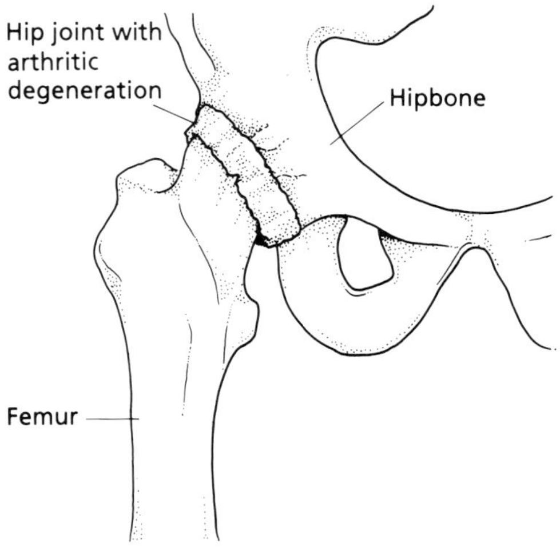

Joint replacement

Originally surgeons either relined the socket in the hip or reshaped the head of the femur. Later developments enabled the head of the femur to be replaced by one of plastic, but the problem of lubricating the points where artificial and natural materials met was never really resolved satisfactorily. Then it was found that, if the head of the femur were replaced by one of stainless steel and the socket by a specially shaped cup made of high-density polyethylene, the problems of lubrication and wear were much reduced. Today, this replacement is commonplace in an operation which has a very high success rate.

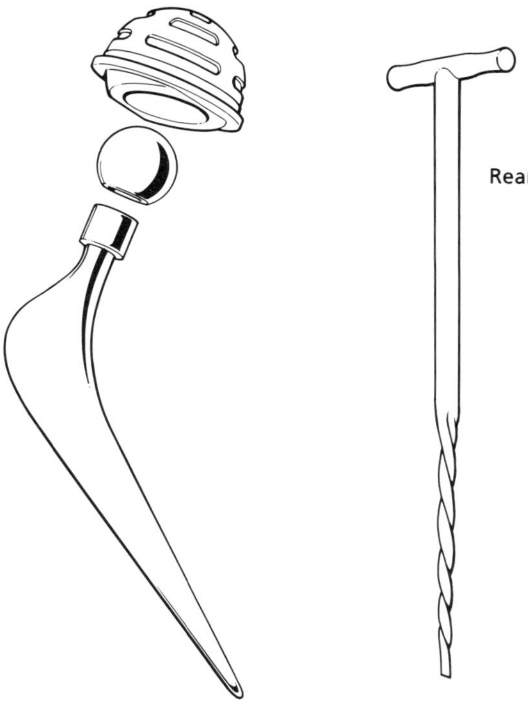

The operation

I have watched several of these operations and have been highly impressed by the technical skill of the surgeon, though the process reminds me more of carpentry than what I had imagined surgical operations would be like.

The operation can take from one to three hours, with the patient under a general anaesthetic, for the natural hip joint to be replaced by a metal 'ball' and plastic 'socket'. First, a long cut is made on the outside of the thigh to reveal the ball-and-socket joint. This is dislocated so that the head of the femur can be removed, using a saw, and discarded. Then the surgeon moves to the hip. The hollow in the hip bone that is to receive the new plastic socket is cleaned out and reshaped so that the plastic replacement can be fixed firmly into place with a bone cement. Next, the metal replacement for the head of the femur has to be inserted. The central core of the femur below where the head has been removed must be reamed out or enlarged and the remaining hole shaped to fit the shaft attached to the metal ball. The ball is then correctly aligned and cemented firmly in position using the important bone cement that we are currently researching into.

After the operation a patient will be testing the new hip in about four days, and will probably discard even a walking stick within three months.

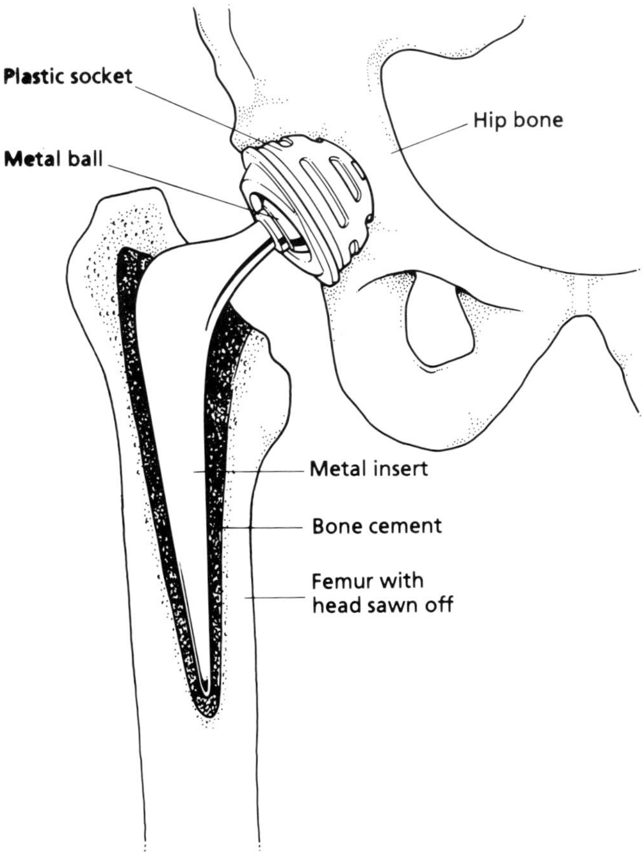

SOME ASSIGNMENTS FOR YOU

This operation is necessary only in extreme cases of joint injury but many people, especially the elderly, are severely handicapped in other ways. I will suggest two possibilities of where your design and technological skill could be useful in 'helping the handicapped'.

The Technician's first project – the design of a telephone dialling aid

Gwendoline Moore is 87 years old. She has led an active life, being employed in a university for many years. She had responsibility for a considerable number of staff and had to make major decisions about research that affected the distribution of Government grants.

Unfortunately, with age her eyesight has deteriorated to the point where she finds it difficult to use the telephone. She is an independent woman who continues to live on her own and is determined to do so as long as she can cope with everyday chores and activities. The telephone is important because it is her only means of communicating with relatives and friends. It becomes even more vital when she is unwell and needs assistance. The telephone is then her lifeline.

Gwendoline's telephone is the common digital type where the call is made by pressing the buttons on the top of the telephone. Her problem is that she cannot see the numbers clearly and has difficulty dialling the correct number. She cannot always remember which buttons are where, and wants an aid which she could place over the telephone to identify each button and help her to dial the number she wants.

Assignment 1 – the design of a telephone dialling aid

You are a member of the research team that has spoken with Gwendoline about her problem but the only information you have is as described above. Your Research Manager has asked you to look into Gwendoline's difficulty in dialling and come up with ideas. The Manager is expecting you to present those ideas, with the help of a mock-up of your solution, to a Charity Committee that helps the elderly. They will be particularly interested in safe, effective, simple solutions that can be provided cheaply to elderly people.

The Technician's second project – opening doors with arthritic hands

Arthur Dalton is 82 and lives alone in an elderly person's bungalow on a housing estate. His wife died two years ago and since that time the arthritis in his hands has gradually worsened. It is so painful that it restricts the use of his hands and he finds grasping everyday objects difficult.

When Arthur was working on the railways he used his hands a lot, for he was a fine craftsman who took pride in the quality of the things he produced. Now he has difficulty in holding a cup of tea. But his greatest difficulty is in opening the front door of his bungalow. He becomes frustrated when people call and he cannot grasp and turn the doorknob. Arthur likes people to visit because he gets lonely, but the struggle with the door becomes embarrassing for him when they do call.

Your research team has been contacted by one of the organisations which help elderly people because Arthur's problem is shared by many elderly arthritic people. The organisation wants a simple yet effective device that could be used to help people like Arthur turn the knob on the lock of their front door. For although Arthur has produced two solutions to the similar problem of how to turn keys in a lock, his problem with turning the doorknob continues to give him much frustration – and pain!

Your team of researchers has been asked to look into the problem and produce prototype solutions for the Elderly Help Organisation (EHO). The preliminary investigation is being funded by this organisation and continued support from EHO will depend upon the success of the initial work, as measured by your report.

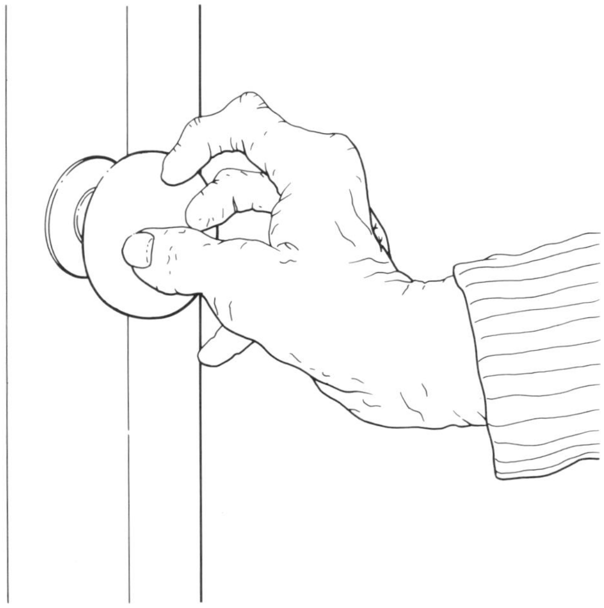

You might begin by considering:

- where you can find information
- how arthritis affects someone's hand
- the different shapes and sizes of front doorknobs/ locks
- a universal device for all types of doorknob
- different devices for different doorknobs

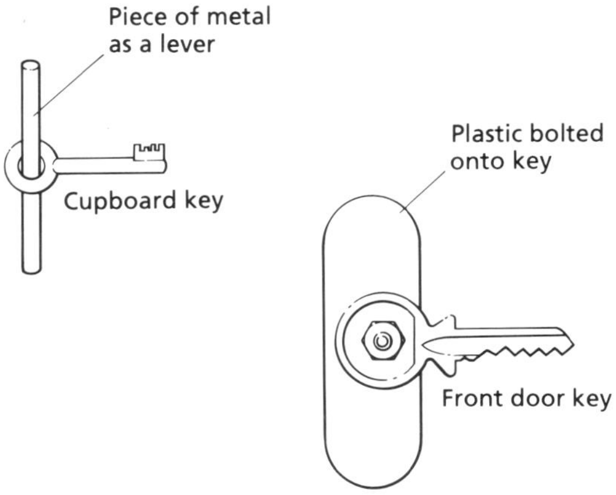

Assignment 2 – designing doorknobs for arthritic hands

Before you start producing your prototype, your Research Manager wants to see the background information you have collected and your initial ideas. These should be presented in an attractive visual form. Your teacher may play the role of this Research Manager.

The interim report is needed for the EHO. Only after seeing this will they decide whether to commit themselves to further funding for the project. In other words, your job is on the line! The EHO will be particularly interested in your estimated cost of the proposed device and the manufacturing system needed to produce it. Its functional qualities are evidently vital. They hope to keep costs down by using a simple, yet efficient, method of production.

Pharmaceuticals – caring for your health

Have you ever thought about the range of medicines that are available in your local chemist shop and wondered where they all come from? The Wellcome Foundation produce a wide range of pharmaceuticals which you can buy at a chemist or have prescribed by your doctor or local hospital.

Because these products are used to help people overcome illness and disease it is very important that they are 100% pure. In the laboratories and production and packaging departments, hygiene standards are very high. All workers have to wear sterile coats and hats and have to change into special shoes when they arrive for work.

Frances Deans works in the mechanised packaging department at the Dartford factory, where she supervises a packaging line. She has six other people working for her and is responsible for the packaging of many types of tablets. All have to be thoroughly inspected and counted. The factory staff must check that there is no contamination, none of the tablets is broken or chipped, and the correct number is in each container.

Two assignments arise out of this work. The first requires the design of disposable clothing for use by those that visit the factory. The second involves a system of counting tablets quickly and efficiently.

BACKGROUND INFORMATION

This information is presented to students in the words of a Packaging Supervisor. In describing the packaging of tablets she indicates that, even though all are packed automatically, there is still a need to check that it is done correctly in a hygienic environment. She describes how the packaging takes place and how the company ensures that packages leaving the factory are properly labelled and contain the correct number of appropriate tablets.

Out of this description arise two assignments, described below, which can be completed independently of one another.

STUDENTS' FIRST ASSIGNMENT AND THE NATIONAL CURRICULUM

Assignment

The consistent requirement throughout a pharmaceutical production process is for the highest hygiene standards. Not only are there primary hygiene barriers regarding personal clothing but for some drugs there are also secondary barriers formed by the automatic handling of the drug in self-contained rooms. This assignment asks students to respond to a real problem that the company has in ensuring that visitors to the plant do not contaminate products unintentionally.

Programme of study

- Take account of human scale and proportion when designing.
- Know that organisations need to have procedures for health and safety, and people responsible for enforcing them.
- Give attention to detail and work accurately.
- Explore a range of solutions before developing one to completion.
- Know that products must allow for unusual eventualities during use.

Attainment Targets

Identifying needs and opportunities (1.5b). Students must 'recognise that economic, social, environmental, and technological considerations and the preferences of users are important in developing opportunities'. ACTIVITY: students design safety clothing and must give consideration to human form and the wearers' tolerance.
Generating a design proposal (2.7b) requires students to 'review the detail of their design, using their own experience and that of others, and suggest alternative ways of achieving what is intended'. ACTIVITY: the best design of clothing is dependent upon an interaction between design ideas and reality. Constant trialling of these ideas and of form are required.
Planning and making (3.8c). Students must 'identify and incorporate modifications during making'. ACTIVITY: the clothing should be of a standard size and yet fit everyone comfortably. Modifications will need to be made to compromise these two design parameters.
Evaluating (4.7a). Students are expected to 'present an evaluation of their activities against the original need, drawing on information gathered about the product and the reactions of users. Evaluation should include suggestions for improvements'. ACTIVITY: students produce a questionnaire for consumer evaluation.

STUDENTS' SECOND ASSIGNMENT AND THE NATIONAL CURRICULUM

Assignment

A typical production problem of sampling is to check a product while the line is in operation. A simple device is envisaged as a solution. This could be a triangular plate to hold sweets, similar to a snooker frame, on which is marked the line for the correct quantity. It is important that the sweets chosen to simulate the tablets are of uniform size and shape. Pharmaceutical tablets should *not* be used. Students will be asked to present their project to a Chief Designer, a role that you might play.

Programme of study

- Recognise and take into account in . . . designing that people can be an element in a system.
- Explore a range of solutions before developing one to completion.
- Design and make equipment to solve particular production problems.
- Present . . . proposals to an audience, using a range of methods and media.

Attainment Targets

Identifying needs and opportunities (1.6c). Students should 'understand how the introduction of new technologies can provide new opportunities and create new demands for design and technological activity'.
ACTIVITY: although the line is fully automatic and theoretically 100% reliable and accurate, the students still have to develop a method for operator sampling.
Generating a design proposal (2.6b) requires students to 'produce a design proposal recording their decisions and the ways of reaching their chosen outcome'.
ACTIVITY: students are asked to submit evidence of their design activities to the Chief Designer.
Planning and making (3.5a). Students should 'identify stages in making and coordinate these into a simple plan to ensure efficient use of time, materials and labour'.
ACTIVITY: there are pressures of time in producing the prototype which have implications for the students' efficient use of that time.
Evaluating (4.5b). Students must 'justify the ideas, materials, components, procedures, techniques and processes used, and indicate possible improvements'.
ACTIVITY: students are asked to present the prototype and supporting material before the Chief Designer to demonstrate its effectiveness.

TABLETS FROM THE BEGINNING

Do you take your health for granted? When you are ill and need medicines and tablets, do you take those for granted too? Only when people find that there are diseases and illnesses for which there is no cure do they begin to appreciate all the work and development that is carried out by industry in finding, testing and producing pharmaceuticals. Then they begin to realise that there is nothing more important than health.

Until quite recently – 150 years ago – there were no cures for many illnesses, and if you caught diphtheria or poliomyelitis or tetanus or whooping cough then the chances were that you would die. Nowadays, there are vaccines to protect you against these diseases and few people in the UK die from them.

Cures and remedies have always been needed to combat illness and soothe the effects of pain and discomfort. If you lived in the middle of the nineteenth century and had an illness or infection. there were pills, powders and ointments but no tablets. If you needed medication, you would probably have visited a local chemist shop and asked the shopkeeper to recommend a treatment. This treatment would depend upon what you could afford and only the wealthy could pay a doctor to diagnose and treat an ailment.

In 1880 two American pharmacists, Silas Burroughs and Henry Wellcome, formed the London partnership of Burroughs Wellcome and Company to sell the new 'compressed' or tablet form of drug that they imported from America. Stamp duty had to be paid on the imported tablets, much to Wellcome's frustration, and so the company started producing its own tablets. By 1889 they had developed special machinery to produce tablets in a standardised form and in a range of dosage from 0.06 mg to 3.60 g. The accuracy of these machines and the purity of the products were extremely high. There were strict laboratory controls of raw materials and the production process.

Henry Wellcome put the words 'tablet' and 'alkaloid' together to form 'Tabloid' as a brand name for his compressed tablets and for many years it was stamped on every tablet. The word is now used to describe 'compressed' newspapers.

Saving lives

The Wellcome Chemical Works started producing pharmaceuticals at Dartford in 1889. In the early days most of the company's products were produced from plants and the Dartford site was used to grow crops of deadly nightshade, foxglove, henbane, thornapple and aconite. The works extracted substances from these plants to use in its products. Digitoxin was extracted from the leaves of the European foxglove for treatment of heart failure and that is still in use today. Nowadays, the number of products made from natural substances is very small, with nearly all the pharmaceuticals produced at the Dartford factory being made from synthetic chemicals.

The Wellcome Foundation employs 19,000 people in more than 50 countries to develop and produce the medicines and tablets that are so vital to our health and well-being. They carry out a lot of research into ways of treating disease and illness. Some of the Company's developments include: diphtheria antitoxin (1894), insulin (1922), diphtheria vaccines (1927–31), yellow fever vaccines (1936), oral polio vaccine (1961) and currently, besides many others, work on the AIDS virus.

Before any new drug can be manufactured, scientists and engineers have to find ways of changing a research chemical into a marketable pharmaceutical. The laboratory methods for producing the new drug are then scaled up into an industrial process.

The first tablets were produced by hand in 1843. Although the basic principles of tablet manufacture are still the same, the quality of tablets has improved greatly. Now tablets are produced using sophisticated computer-controlled machines which eliminate the possibility of contamination and produce tablets of high purity and accurate dosage.

Modern medicines produced by Wellcome include: Imuran, used in transplant surgery; Retrovir, and antiviral treatment for AIDS; and well-known cough and cold treatments.

PACKAGING TABLETS

I'm Frances Dean and I supervise a packaging line at our Dartford factory. In our packaging hall we put tablets in blister packs and in security containers similar to those used for aspirin. We also bottle liquids. All tablets are packed by automatic machinery. We fill each machine with the tablets to be packed and the machines pack them for us. However, they cannot think for themselves and so we have to check that they are operating properly. If a machine jams we need to find the cause of the problem, sort it out and restart the machine. Our aim is perfection and for this we must continually check the blisters and containers. All the spaces in each blister must be filled by a tablet and each container must hold the correct number of tablets.

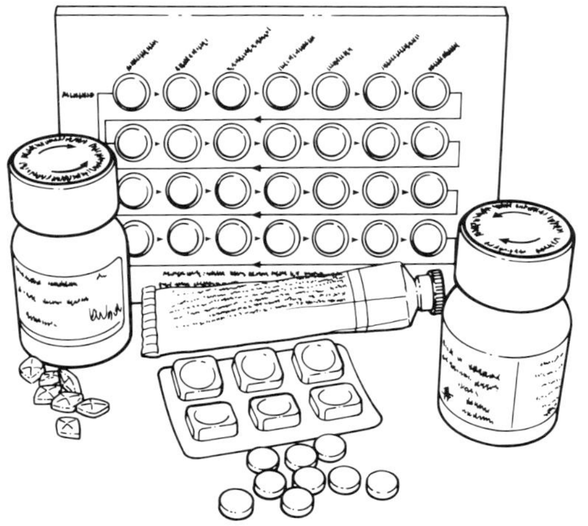

Blister packaging

A blister pack holds each individual tablet separate from all the others. The blister packing machine first feeds tablets into a hopper that passes them down a tube towards a conveyor. The film material which holds the tablets is then fed in and has dents for the tablets pressed into it. Next, the tablets pass a sensor that sees that each is fed into its own dent. Over them a camera is positioned so that one of the operators can check that all the dents are full.

Then the top layer of film is fed in. Printed with the description of the tablets, this film is pressed onto the one that contains the tablets. As a result, a huge series of blister packs is formed. Smaller packs are cut from this before they are placed in cartons for storing in larger, foil-wrapped boxes. Today the machine is producing packages, with five tablets in each blister, at a rate of 350 blisters per minute.

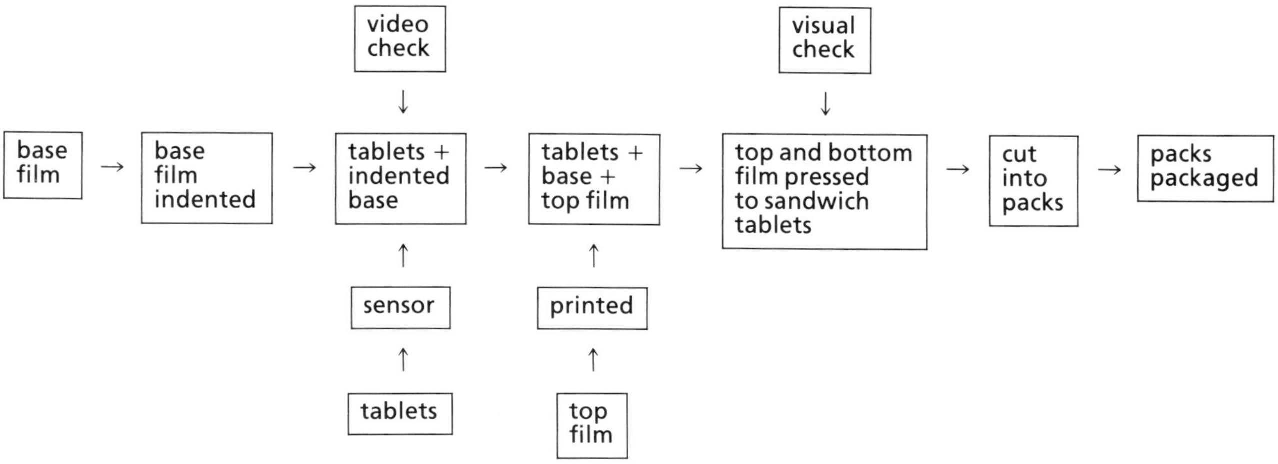

Security containers

Tablets fed in Archimedean screw

A different machine automatically packs into their containers some red, film-coated tablets. We put the containers into a machine which cleans them by blowing in and drawing out air. The tablets are then placed in a hopper and fed to another machine by an Archimedean screw (see above). The machine is set to count 100 tablets and only allows this number to enter each container. It continues filling containers at about 100 per minute. The filled containers travel along a conveyor to a turntable, where they wait for the next stage of the packaging process. We sample the contents of a container to see whether the correct quantity is still being delivered by the machine to each container.

When the machine is ready, it allows the filled containers from the store to continue along the conveyor to the next stage. Normally, this would put cotton wool wads into the top of the container to stop the tablets hitting each other in transit and chipping or breaking. However, because these tablets are film-coated they are much harder and don't need such protection. The machine next feeds in the container-tops and presses them into position on each container. If by any chance the machine misses a container, one of the operators steps in and presses on a top.

The description of the type of tablet and dosage instructions are very important. Also important is the 'lot number', which tells the consumer when the batch of tablets was packaged. All this information is printed on adhesive labels, placed by the machine on every container. The containers are now placed in trays and vacuum-wrapped to stop dust and other small particles contaminating them. Quality assurance tests are then carried out before the goods are released for shipment.

The Packaging Supervisor's first project – the design of hygienic disposable clothing for visitors

Because we manufacture tablets for people suffering from a variety of illnesses or ailments, we have to ensure that all our production and packaging processes are completely hygienic. To prevent contamination by personnel, each individual wears special footwear, headgear, and a laboratory coat. This clothing is put on before entering the production and packaging areas.

Visitors often call to inspect our processes or to talk with our staff. We give them a laboratory coat to wear, and also covers for their hair and shoes. Because these items will be used only once, they must be cheap. One head cover must fit all head sizes and similarly one shoe cover must fit all shoe sizes. It is important that the covers are comfortable and do not slip off.

Perhaps you could help us by designing and producing prototype covers to meet our requirements. You could produce your covers from paper such as paper towelling, or even kitchen cloth. Make a verbal report to your teacher of your recommendations.

Assignment 1 – designing disposable clothing
Draw up a design specification for the covers, both head and shoe, using the information given here. Remember, the main purpose of the covers is to prevent contamination of our product.

Produce alternative designs and compare the advantages and disadvantages of each one. Test your final products by asking others to wear them. Produce a simple questionnaire for each wearer so that you can report which design is best.

The Packaging Supervisor's second project – a device for counting the tablets in each container

When we package our tablets, it is very important that the correct number is placed in each container. You have probably seen bottles or plastic containers with the quantity of tablets written on a label which is stuck to the outside.

Our automatic packaging machines inspect the tablets for faults, then count and put them into containers. However, to make sure that each machine is operating correctly and putting the correct quantity of tablets in each jar, we manually sample the contents every five minutes. We cannot do this less frequently because the machine packs 100 containers every minute and if it were making a mistake we would have a lot of incorrectly filled containers.

If the operator were to count the tablets one by one, it would be a laborious and time-consuming task. In addition the operator could make mistakes with counting, especially when some containers contain 100 tablets. We want some way of visually checking each container for the correct number of tablets which is quick, easy for the operator to complete, and eliminates the possibility of any mistakes occurring.

Our design team has started work on this problem and has considered putting the tablets from the container into a scoop or a funnel for counting. This could be held by the operator while the quantity is read off. One scoop/funnel would be made for each type of tablet produced. We would like you to continue with this line of enquiry but use small tablet-sized sweets instead of tablets. DO NOT USE PHARMACEUTICAL TABLETS.

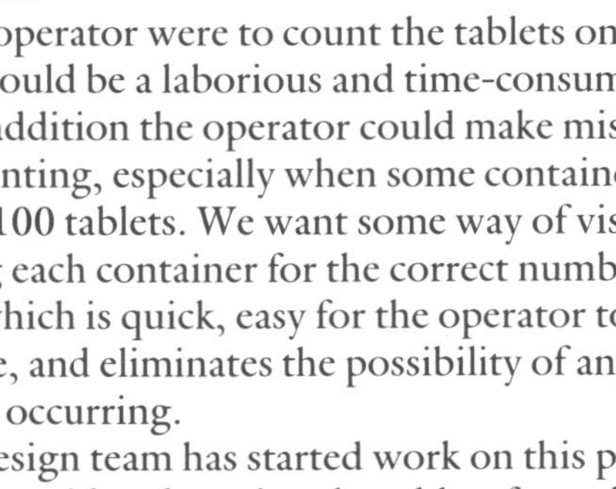

The scoop or funnel must give the operator an immediate visual check on the quantity in the container without counting every single tablet or, better still, without counting any of the tablets!

Quick

Reliable

Avoids counting

Visual

Easy to use

Accurate

Hygienic

Assignment 2 – counting tablets

Imagine you are a member of the design team working on this problem. The team has been asked to report to the Chief Designer as soon as there is a working prototype.

The Chief Designer (a role that may be played by your teacher) wants to see a report of the team's design activity which includes:

- sketches of all the team's ideas for solving the problem
- a materials specification
- estimated costs for producing the prototype
- detailed drawings of the prototype
- a breakdown of the design work showing how much time each stage took before a finished prototype was made.

In addition, the Chief Designer wants to see an instruction manual for informing someone how they should use the device. She will expect to see a demonstration of the prototype showing how it is to be used.

Weaving – carpets for ever

Some carpets last longer than others. Some fade and lose their colour; others easily become permanently stained; and yet others lose their pile or just wear away. The way in which a carpet is designed and produced is vital to long and economical use.

Weaving carpets involves a technology that has developed over thousands of years. Now it is so advanced that, unlike thirty years ago, many people in the UK have wall-to-wall carpets in their homes.

Wilton Royal makes carpets in Wiltshire and Debbie Hunt is a designer. Debbie will explain what her job involves and provide two assignments. The first of these requires the discovery by survey of the factors carpet-buyers consider important. The second requires the design of a suitable carpet for a school entrance.

BACKGROUND INFORMATION

Information about the Wilton Royal carpet factory and the work done there is given in the Student Materials. This includes a brief description of how carpets are now designed and produced, and how fibres are chosen, dyed and woven. The description leads to the presentation of two assignments which are independent of one another.

STUDENTS' FIRST ASSIGNMENT AND THE NATIONAL CURRICULUM

Assignment

A carpet manufacturer needs to identify and place in order of priority the factors that customers consider important when buying carpets for particular purposes.

Programme of study

Review the ways in which market research can be used to evaluate user requirements and market potential.

Attainment Targets

Identifying needs and opportunities (1.10a). Students should 'elicit and interpret the perceptions, motivations and needs of people in a range of contrasting situations'. ACTIVITY: information is collected by direct interview, the form of this interview to be decided by the student groups.
Generating a design proposal (2.7c) requires that students 'apply relevant criteria including user requirements and costs'. ACTIVITY: criteria investigated regarding carpet usage include cost, wear, dirt resistance, flammability, and fade.
Planning and making (3.7a). Students should 'plan and carry out working procedures to match the constraints on making'. ACTIVITY: a series of decisions has to be made in planning the survey.
Evaluating (4.7a). Students are to 'present an evaluation of their activities against the original need . . . including suggestions for improvement.' ACTIVITY: the required report has to include an evaluation of the survey with comments on how it could have been improved.

STUDENTS' SECOND ASSIGNMENT AND THE NATIONAL CURRICULUM

Assignment

A manufacturer of carpets has had a request from a school to design the floor covering for the school's entrance hall.

To gain experience of carpets, a brief carpet survey should be carried out with the help of a local shop or mail order catalogue. A complete entrance hall has to be designed and a card model produced of the design, with perhaps a piece of coloured paper to represent the carpet itself.

Programme of study

- Design the appearance of an artefact, system or environment so that it appeals to users.
- Use modelling techniques to convey a design proposal.

Attainment Targets

Identifying needs and opportunities (1.6c). Students should 'understand how the introduction of new technologies can offer new opportunities'. ACTIVITY: the use of a computer-assisted Jacquard loom provides the opportunity to include a school logo or school colours in the carpet being designed.
Generating a design proposal (2.6c) requires students to 'make judgements about realistic ways forward by exploring alternative solutions and use these to refine their design proposal'. ACTIVITY: the activity includes producing a cardboard model of the entrance hall so that different designs can be evaluated and presented for inspection.
Planning and making (3.6f). Students should 'use knowledge of technical and symbolic representations of materials, components'. ACTIVITY: the model is considered an example of a symbolic representation.
Evaluating (4.8b). Students should 'understand that artefacts, systems or environments reflect the circumstances and values of particular cultures and communities'. ACTIVITY: the model could be taken to the headteacher for comment and instant evaluation as to how the design reflects the school's ethos.

DESIGNING WITH TECHNOLOGY

I'm Debbie Hunt, a Carpet Designer. When my office phone rang this morning, it was a call from the owner of a large hotel in Torquay who needs to replace the carpets in each of the hotel's 150 bedrooms. She asked if I could arrange to visit her hotel to meet the manager, and explain the services offered by Wilton Royal.

During my visit I must get the owner to show me the hotel and its bedrooms and so find out what she believes is required. I need to know:

- Is the hotel modern?
- Does it have a particular style?
- Do all the bedrooms have similar decor?
- Is the hotel's general decor to be renewed with the carpets?
- Are the bedrooms all the same size? What are their sizes?
- How many bedrooms have *en suite* bathrooms?
- Does the hotel have a logo or hotel 'colours'?

After the visit I will produce several designs, using a computer to help me. As you know, a computer is only a tool. If I put in poor ideas, it will give me back bad designs – but quickly! When designing a carpet, and with a basic pattern on the screen, I can change colours and shapes immediately. The days of drawing and colouring my designs by hand on paper have gone, thank goodness.

Clever as the computer is, it cannot produce the exact colour of the carpet. This can only be shown by reference to the dyed materials, so I have small samples of carpet made up by hand and take them to potential customers as soon as they are available.

The first impression gained by the hotelier will be of colour, pattern and texture. These are of considerable importance as at this stage I will not have made a sale and it is still possible that the hotelier will reject my designs and go elsewhere.

Choosing fibres

My customer, the hotel manager, will not be pleased if, after a few months, the carpets that once felt wonderful to walk on now show pathways in their pile where people have trodden. However, there is more to carpet quality than just not wearing out. The manager will also be dismayed if the carpets, while 'lasting for ever', are unpleasant to walk on, or if static electricity gives guests a shock and attracts dirt, or if the guests accidentally set the carpets alight with burning cigarettes. Ease of cleaning is another factor I must keep in mind. Some cheaper fibres that are used for making carpets are easily stained and extremely difficult to clean.

The table may help you see the advantages and disadvantages of just a few fibres. It is a simplified picture as the woven material's properties can be altered by weaving in different fibres. As a result, many carpets are made from mixtures of fibres that give a balance of properties.

	Wool (natural)	Nylon (synthetic)	Acrylic (synthetic)	Polyester (synthetic)
Wear	Good	Very good	Good	Good
Crush	Good	Exceptional	Good	Good
Fire	Very good	Good	Poor	Good
Dirt	Fair	Average	Good	Good
Grease	Good	Poor	Poor	Poor
Static	Good	Poor	Poor	Poor

Key
Exceptional (6 dots) Good (4 dots) Fair (2 dots)
Very good (5 dots) Average (3 dots) Poor (1 dot)

As you would expect, cost is very important. Wool is far more expensive than a synthetic fibre, but it also has that indefinable quality of 'feel'.

Dyeing fibres

Assuming that one of my designs has been accepted, the price agreed and an order placed, we can begin to make this carpet as a special order.

One problem is that after placing an order our customers want their carpets quickly and on time. Manufacturing a carpet needs a great deal of effort by many people and so we produce a job-sheet that specifies the date for each task's completion. This coordination ensures we know where we are and can still deliver as agreed.

For this Wilton carpet I will recommend a mixture of fibres, 80% wool and 20% nylon. This mixture will give the carpet a balance of properties that suit the hotel's bedrooms and its general decor.

We buy our synthetic fibres from a company in Bradford. As you may be aware, the basic raw material for this fibre is crude oil. The natural fibre, wool, arrives in its raw state, having been shorn from New Zealand sheep. It needs to be washed several times, first to remove grease and then to remove dirt. Afterwards, it looks white and fresh but, of course, is the wrong colour for my designs.

Before weaving the fibres, both the cleaned wool and the synthetic fibre must be the colours specified in my original design. Different materials need different types of dyes. Once natural dyes were used but today synthetics are for sale in many more colours and shades than natural dyes are. Also, synthetic dyes do not wash out: they are colour fast.

The colours obtained with synthetic dyes can, with care, be reproduced time and time again. As there are so many colours and shades available I could, when making the original design, match the colour of the carpet with the colours already present in the hotel's bedrooms or to be introduced into the hotel. Whatever the choice, the vat containing the dye must be controlled carefully for acidity, temperature and the time taken to dye the fibres.

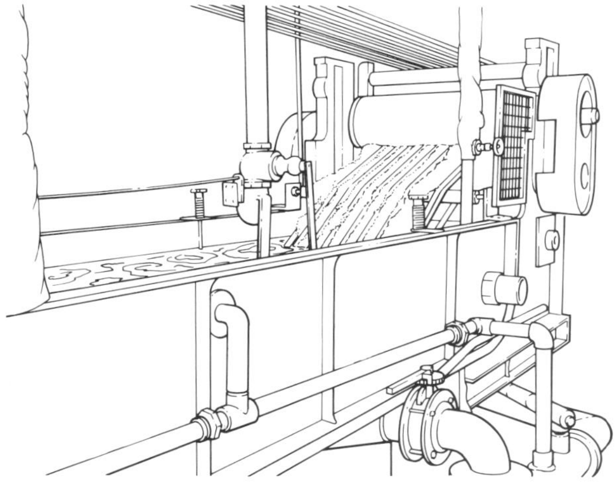

Simple weaving

Before I can explain how weaving takes place at Wilton Royal, let's look at basics. To weave a fabric, a row of threads – called warp threads – are held lengthways. At right angles to these warp threads, weft threads are interwoven one at a time.

The type of loom illustrated, used by ancient Egyptians, weaves quickly. A shuttle holds the unwoven fibre. Can you spot the batten that is used to push the wefts together, so forming a dense weave?

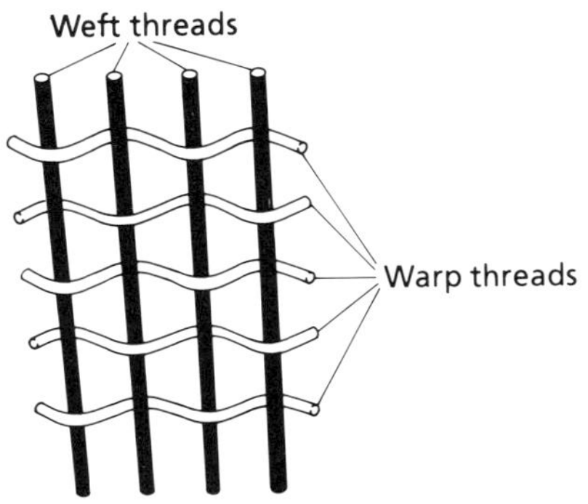

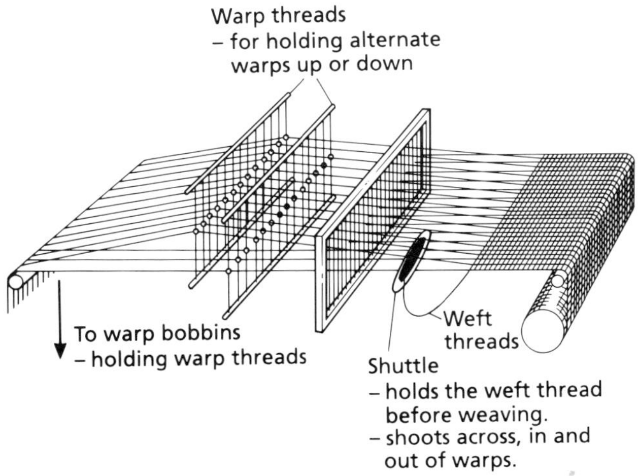

If different coloured warps and wefts are used and there is some way of lifting the warp threads in different orders, the woven material can have a coloured pattern. But, with this method, there is no pile.

Such a simple loom would not produce the complex weaving we require at Wilton. We use Jacquard looms, with holes in punched cards that allow rods to lift the yarns to form loops, the 'pile' of the carpet. When the loops are cut, as for the hotel's Wilton carpet, the result is a cut pile. The pile is sheared after the carpet has left the loom and as this pile has to be worn out before the base of the carpet is reached, the carpet is extremely hardwearing.

Finally, the back of the carpet is strengthened by coating it with a pliable, synthetic material. Keeping to our job-sheet will ensure that all tasks are coordinated so that the carpet is delivered on time. We hope that one day the hotel manager will order another carpet, perhaps to fit the hotel's reception area.

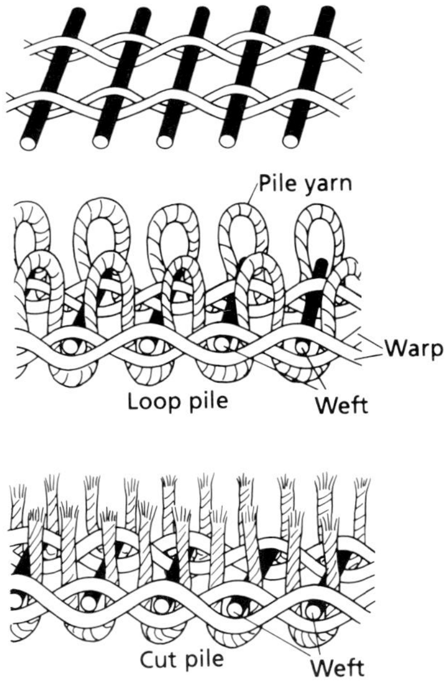

The Designer's first project – factors considered important when buying carpets

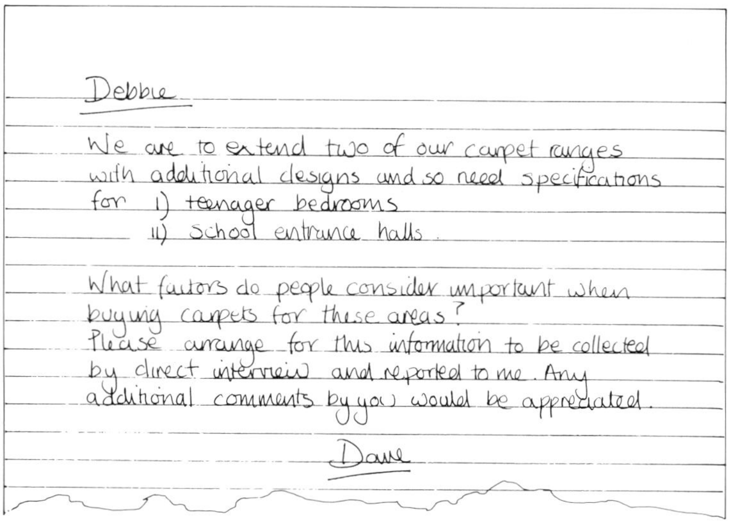
Debbie

We are to extend two of our carpet ranges with additional designs and so need specifications for i) teenager bedrooms
ii) school entrance halls.

What factors do people consider important when buying carpets for these areas?
Please arrange for this information to be collected by direct interview and reported to me. Any additional comments by you would be appreciated.

Dave

I've just received a memo from my Production Manager. I need your help to collect the information required. We consider these factors important but which are the most important?

cost *wear*

crushability *pattern*

colour *feel* *flammability*

dirt resistance

static electricity

washability

grease resistance

fade

Assignment 1 – factors considered important when buying carpets

1 Work in a small group to plan the investigation but divide up the tasks between yourselves after answering these questions:

- Who should be interviewed?
- Who should do the interviewing?
- What should be asked?
- How should they be asked?
- How should records be kept?
- How should the results be analysed?
- How should the results be presented?

2 Complete the investigation.

3 Report by:

- explaining how you have completed your survey
- listing separately the factors, in order of importance, that people use to choose carpets for each of the areas
- giving your own evaluation of this report's accuracy and how the survey could have been improved by including methods other than direct interviews.

The Designer's second project – carpets for school reception areas

A report about successful hotels stated: 'As you would expect, we like to create a good initial impression with our visitors. One way of doing this is to make the reception areas as attractive and as comfortable as possible. The decor should reflect the ethos of the establishment. An extreme example of failure to do this would be to have photographs of a hunting scene in an RSPCA reception area.'

I must admit that I'm influenced by the reception area of a school when I first enter. It can create a good, or a bad, impression. Perhaps I am biased, but I always look at the quality, the design and the colour of the decor. Of course, some schools make serious mistakes in their choice of floor covering, and only some take advice from designers like myself.

Completing a carpet survey

Before you can help me with this project you will need to have more practical experience of carpets. To gain this, you will need to visit a local shop that sells carpets or, if this is impossible, use a mail order catalogue.

If you go to a shop, ask whether you can complete a simple carpet survey. If the shop agrees, choose several carpets and complete a table something like the one shown.

Name of carpet
Type of pile (density and height)
Cost (per metre)
Durability (number of years)
Composition (e.g. 80% wool 20% nylon)
Type of underlay
Cost of underlay (per metre)
Widths sold (metres)

Once you have a little more experience with carpets, you will be able to help me with my next problem. A school has written to me about their entrance hall. By a great coincidence, it has exactly the same shape as the entrance hall to your school.

I have been asked to create a 'good' impression. The headteacher insists that the floor is carpeted. Your task is to make the first attempt at designing and producing the specification for this carpet. Of course, the design cannot be done in isolation, and you will have to make recommendations for the other decorations within the entrance, including any additional furniture that is required.

Certainly you will have to decide what image your school is trying to create for itself. Are some of these suitable?

industriousness

prize winning

cooperation

cleanliness

thriftiness

wealth

academic excellence

a caring community

happiness

sporting prowess

Assignment 2 – carpets for school entrances

1 Decide what image your school wishes to convey.

2 What wall covering, curtains, and room layout will convey this image?

3 Produce a design, perhaps using computer assistance, and specifications for a carpet. Don't forget the experience you gained in completing the survey of carpets that are for sale, and the factors people consider are important in carpets.

4 Consider where, if anywhere, the school's emblem/badge/logo will appear. Could it be in the carpet?

5 Test your designs by producing and inspecting a cardboard model of the reception area and position the carpet (a piece of coloured paper will do) in the model.

6 Don't forget that an estimate of costs is important.

7 You might then ask to see your headteacher. Show her/him your plans and see whether she/he believes they reflect the values of your school.